Plastics
and
Sustainable Piping
Systems

David A. Chasis

Industrial Press

A full catalog record for this book will be available
from the Library of Congress
ISBN 978-0-8311-3498-3

Industrial Press, Inc.
32 Haviland Street, Suite 3
South Norwalk, Connecticut 06854

Sponsoring Editor: John Carleo
Interior Text and Cover Design: Janet Romano
Developmental Editor: Robert Weinstein

1 2 3 4 5 6 7 8 9 10

TABLE OF CONTENTS

TABLE OF CONTENTS BY TOPIC

ACKNOWLEDGMENTS

I would like to acknowledge and thank two industry associations who have been critical in the birth of this publication: International Association of Plastic Distribution (IAPD) and Plastic Pipe and Fittings Association (PPFA).

The IAPD is an international association established in 1956 comprised of plastic industry distributors and companies. Their well-known publication, *IAPD Magazine*, is a bimonthly journal which has been a positive force in educating distributors, engineers, installers, and end-users to the features and benefits of plastics. Many of the published articles appearing in the book were first printed in the *IAPD Magazine* as noted. I would like to thank IAPD for giving me permission to include these articles in the publication. Please go to www.iapd.org website for more information on IAPD.

PPFA, an association in which I am a member and a consultant, has been very generous and helpful by being a source of information for many of the articles and photographs appearing in the book. Many of the article drafts have been peered reviewed and greatly improved by the input of PPFA members. There is no association in North America that has done more in getting plastic piping systems accepted and approved in residential and commercial building codes than PPFA. PPFA is also a stalwart force in promoting and defending the use of all thermoplastic piping in the United States and Canada. Please go to www.ppfa-home.org website for more information on PPFA.

The article "Plastics Learn Their Roles in Plumbing System Design" has been published in the American

Society of Plumbing Engineers'(ASPE) journal *Plumbing Systems and Design* (journal has since been renamed: *Plumbing Engineer*). ASPE has granted permission to reprint the article in the book. Also, the article "Think Plastics: Thermoplastic Piping Gaining Industry Acceptance" was published in the Valve Manufacturers Association's (VMA) journal, *Valve Magazine*. The VMA has granted permission to reprint the article in the book. My thanks to both of these well-recognized and influential professional associations.

The author would also like to recognize the many companies whose associates have been very helpful in contributing photographs and copy edits for many of the included articles. The companies are:

- Aquatherm, Inc.
- Arkema, Inc.
- Asahi-America, Inc.
- Charlotte Pipe and Foundry Company
- Chasis Consulting, Inc.
- George Fischer LLC
- George Fischer Harvel
- Hayward Flow Control Systems
- Indelco Plastics Corporation
- IPEX Industrial Systems
- IPS Corporation
- JM Eagle Corporation
- Kaneka Texas Corporation
- LASCO Fittings, Inc.

- Lubrizol Corporation

- Mueller Industries, Inc.

- NIBCO INC.

- Oatey Company

- Rocky Mountain Colby Pipe Company

- Shintech, Inc.

- Uponor

- Victaulic Company

- Wavin N.V.

- Zurn Industries

Much gratitude goes to proof readers, Word savants, and editors, Adrienne Travis, Arvis Green, and Robert Weinstein. I would also like to recognize the many creative and insightful contributions of Janet Romano and John Carleo of Industrial Press, Inc. Thanks also goes to the book's marketing strategists Christine Ott and Lisa Flournoy Richardson. The input of this team of talented professionals most definitely improved the format, grammatical correctness, and clarity of the presented articles as well as helped create a larger audience and more user friendly publication. But please note that any error that may appear in this publication solely rests with the author.

Last, I am greatly indebted to many of the pioneers in the plastic piping industry—especially my late and close friend, Dennis Garber, who was the most passionate person I have ever known regarding the wonders of plastic piping. He was one of the first plastic piping zealots.

PREFACE

Several years ago I was making a sales call on a consulting engineer who specialized in power plant design. Before the meeting, his secretary mentioned that the engineer was hospitalized recently with heart arrhythmias and had a pacemaker implanted. During my sales call, I presented the features and benefits of plastic piping to the engineer. After a lengthy and unsuccessful attempt to convince him to consider plastic piping for intake and outtake lines, as well as water treatment systems, he blew me off; he stated that plastics were inferior to metals and were too "flimsy" and "unproven."

As I left his office, I mentioned his recent implant. I asked him "Do you know what material was used in the manufacture of the device housing that is keeping you alive?" "Plastics!" I said. He never answered and just angrily shooed me out of his office. The point being: too many engineers are mired in old technologies; they are not open to or aware of new and better construction materials. The end result is that the engineer's clients are subjected to less effective and more costly construction.

Plastic is defined as a synthetic material made from a wide range of organic polymers that can be molded into shape while soft, and then set into a rigid or slightly elastic form (definition adapted from Oxford English dictionary). These man-made materials have grown astronomically over the last 100 years, offering many features and benefits throughout such industries as clothing, appliances, transportation, packaging, chemical processing, electronics, and medical. But no industry has benefited more from the advent of plastics than that of construction, building with such products as wall and floor coverings, window frames, siding, roofing, decking, fences, and piping systems. This book focuses mainly on plastic piping systems.

I was introduced into the plastic piping industry when I moved to Louisville, Kentucky, in 1967 to work for Cabot Piping Systems (now the Chemtrol division of NIBCO INC). My official job title was marketing manager, yet my duties included sales, distribution, product development, and any other quasi-marketing job that was sloughed off for me to finish. Thanks to the cooperation of my Cabot associates and the company's extensive product line, I took to plastics as a duck takes to water.

And for the next forty-five plus years, I have been involved in all facets of the plastic fluid handling product industry including: marketing, manufacturing, distribution, engineering design, customer service, domestic / international product sales, corporate acquisitions, and consulting.

In the late 1960s and early 1970s there was very little published educational information available to engineers, installers, and end-users in the use of plastic piping systems. So I undertook a part-time two-year process to create the "bible" of plastic piping and voilà in 1976 *Plastic Piping Systems* was published. In 1988, the revised edition of *Plastic Piping Systems* came out. Both books were published by Industrial Press, Inc.

In the past 25 years, I have authored four dozen articles on plastics, many of which have been published in industry journals. In addition to printed media, I have been involved in creating and presenting dozens of seminars, webinars, and tutorials to hundreds of curious and open-minded engineers, contractors and large end-users. In 2010, I also created a website to offer free internet access to users who want the latest in plastic piping information:
www.sustainablepipingsystems.com

The idea for this book came to me when there were over 12,000 visits last year to my website, most requesting information on plastic piping. I thought that if I could edit and assemble many of the articles I wrote in a one-source document, it would benefit those who want to find answers to their questions regarding plastics and plastic piping systems—hence, the creation of *Plastics and Sustainable Piping Systems.*

The book contains 46 articles divided into three sections. The first section, *General Plastics*, comprises articles that address the present, past, and future of plastics in general. Many of the articles explain, defend, and promote the use of plastics in the world, offering several examples of why plastics should be the environmentalist's best friend.

The second section, *General Plastic Piping*, concentrates on plastic piping areas such as applications, joining methods, and the environment. Almost all the piping materials listed are thermoplastic (plastics that can be easily recycled and transformed by applying heat and then cooled with the end product having the same physical properties of the original material). (There is no coverage in the book of thermoset piping materials such as fiberglass reinforced polymers. Thermosets are plastics that, once formed, cannot be recycled back into its original form.)

The final section, *Plastic Piping Materials and Products*, focuses on particular fluid-handling product groups such as valves, pipes, fittings, and fabrications as well as product-specific piping material systems. These articles are more likely to answer many of the questions the reader may have on a particular plastic piping product or system.

Several of the published articles have been re-edited to deliver more grammatically and technically correct information. Also, charts and photos have been added to some of the articles to better improve the reader's knowledge and understanding of the presented data.

Except where noted, the information presented in the articles is the sole responsibility of the author and does not necessarily represent the views of any other author, company, or industry association.

David A. Chasis
Austin, Texas
May 2014

PART 1

General Plastics

GENERAL PLASTICS

The following articles promote the benefits of plastics by presenting current and historical facts as well as anecdotal cases:

1. *Celebrating 100 Years and Going Strong* lists the many positive changes plastics have brought to the world after the first all-synthetic plastic was introduced in 1907 by Leo Baekeland.

2. *Chemicals and Plastics* makes a solid case for the concerns addressed by the chemical and plastics industry in following its mandate to provide tested and safe products to the public.

3. *The China Study* article highlights aspects of the comprehensive scientific study done by T. Colin Campbell and his son in China, showing that lifestyle and diet — and *not* man-made chemicals — are the largest contributors to human disease and poor health.

4. *Defining Bioplastic Terminology* explains the many material terms used in industry that begin with the prefix "bio."

5. *The History of North American Plastic Piping Distributors* informs the reader how the phenomenon of special plastic industrial distributors grew the market for plastic piping materials in North America.

6. *Internet...Cure or Blessing for Plastics* outlines tools and strategies for using the Internet to promote and defend plastics.

7. *Pearls of Plastic* addresses an anti-plastic article authored by Brian Walsh in his 2010 article published in the magazine *Time*.

8. *Plastic Piping Systems and the Internet* lists and describes several helpful websites that furnish user-friendly, up-to-date, accurate information on plastic piping systems.

9. *Plastics and PETA* summarizes how plastics have contributed in preventing the slaughter of hundreds of thousands of animals.

10. *Potable Water for a Honduran Village* describes how a group of profit and not-for-profit organizations worked together to provide potable water for a small village in Honduras.

11. *PVC Pipe Used for Non-Piping Applications* discusses and displays more than a dozen examples of PVC pipe being used to build decorative products, furniture, storage solutions, recreational games, and other creative products.

12. *Transformation from a Radical to a Rational Environmentalist* is a review of the book *Confessions of a Greenpeace Dropout: The Making of a Sensible Environmentalist* by Patrick Moore, co-founder and former activist of Greenpeace.

13. *Vinyl and the Planet of the APES* exposes the pseudo-scientific attacks against plastics from a group labeled by the author as the Anti-Progressive Extremists Society or APES.

1

CELEBRATING 100 YEARS AND GOING STRONG

Did you know that the very earliest plastics were made in the mid 1800s from such biodiverse products as natural rubber (rubber tree), gutta percha (balata gum tree), shellac (lac insect), casein (cow protein), bois durci (animal blood and wood), and celluloses (cotton and other plants)? In 1907, Leo H. Baekeland, a Belgian-American chemist, created from phenol formaldehyde the first all synthetic plastic. The invention of this phenolic thermoset compound named after him, Bakelite, was credited as ushering in the age of modern plastics 100 years ago. And what an age it has been!

It's hard to think of any segment of the developed world where plastics have not had a profound impact. From stealth bombers to pacemakers, from countertops to stockings, from flexible electronic circuitry to pipe, from carpets to packaging—plastics have changed the world in which we live. And for the better!

The environmentalist and animal activists should be the first to recognize and embrace what plastics have done to preserve our planet. Just think, the first group of bio-plastics was invented as substitutes for whale bone, ivory tusks, and exotic woods. With the advent of synthetic fibers such as nylon, polyester, and polypropylene, fields of cotton and flocks of sheep and goats were reduced, causing less land pollution and deforestation of the countryside. The need for cattle hides was greatly reduced with plastics replacing leather golf balls, football helmets, upholstery, shoes, coats, and luggage. And don't forget the significant savings of endangered animals whose furs used in winter wear were emulated by par excellence copies from plastic-made look-a-likes. Trees, too, were spared when plastics started replacing paper and cardboard products in packaging and wood used in boat construction, decks, fencing, furniture, windows, doors, shingles, siding, and even replacing wooden pipe.

Figure 1
Synthetic hides
save animal lives.

Figure 2
Ivory replacement also
saves animal lives.

Another saving grace to the environment was when plastics replaced lead-based products. Just in the last few decades, copper piping was joined using a lead-based solder, which at times allowed lead to be leached into the drinking supply at unacceptable and dangerous levels. This same leaching result was seen in brass-lead alloys used in manufacturing water valves and faucets coming in contact with potable water. Plastics became an acceptable alternate product in these applications. Do you remember lead-pigmented paint and the scare it caused? Again, plastics came to the rescue with plastic pigments (acrylics mostly) replacing lead.

But probably the greatest legacy to date of plastics is its contribution to the incalculable savings of the world-wide use of energy. How? Due to plastic's amazing characteristic of strength-to-weight-ratio. Look around us and see what strong, lightweight polymeric materials have done. For example, most of today's automobiles are built with 15 to 25% of their construction in plastics, greatly reducing the vehicle's weight and, hence, increasing the miles per hour fuel efficiency.

Figure 3 Sanding a plastic auto bumper.

The same is true in the design of boats and planes. Then there is the energy savings used for transporting plastic products that have replaced other bulky heavy weighted non-plastic materials. Think of the many more lengths of plastic pipe that can be shipped per carrier versus other products. How about the transportation savings in plastic packaging versus glass containers? In this instance not only is there savings in shipping but also there is considerably less breakage as well as less personal injuries using plastics.

Plastics are so ubiquitous and prevalent in the world that company product trade names have become generic in nature—Tupperware (polyolefins), Cellophane (cellulose), Scotch Tape (cellulose), Plexiglas (acrylic), Lucite (acrylic), Nylon (polyamide),Teflon (polytetrafluoroethylene), Lexan (polycarbonate) and Rayon (cellulose) just to name a few.

What's in store for the next hundred years in plastics is anybody's guess, but you can be sure that it will be life-changing. With the advent of nanotechnology and organic chemistry research advances, plastics will continue to offer technology solutions to many of the problems affecting our planet. And with the importance and focus on sustainability and greenness, it wouldn't be surprising that we come full circle, with plastics that replaced expensive and unsustainable materials in the mid 19th century being themselves replaced with new biofuel-produced plastics that will impart a gentler footprint on the earth.

Reprinted with permission of the IAPD; issue june/july 2007 — **the IAPD Magazine**

CHEMICALS AND PLASTICS

All earthly matter is composed of chemical elements that have a particular composition, structure, properties and reactions, especially of atomic and molecular systems.[1] As of 2010, there are 118 elements as listed in the Periodic Table, with 94 appearing naturally and the remainder created artificially by man-made particle accelerators.

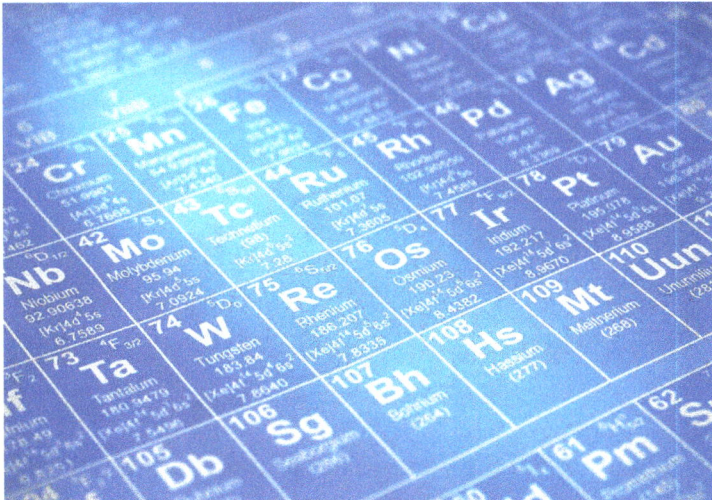

Figure 1 Periodic chart.

The human body is comprised of 59 natural elements with over 99% of its mass consisting of 10 elements in descending order of volume: oxygen, carbon, hydrogen, nitrogen, calcium, phosphorus, sulfur, sodium and chlorine (approximately ninety-eight of body mass consists of the first 6 elements listed). The remaining 49 elements are present in very small or trace quantities.[2]

Even though scientists have been exploring and probing into the chemical make-up of the human body for decades, there are several trace elements such as rubidium, silicon, vanadium, and several others that have yet to be determined what role, if any, they play in the body's metabolic or biologic system.[3] In other words, we have natural chemicals in our body that no one yet has determined what purpose, if any, they provide.

Figure 2 Research laboratory.

Be aware that the body's chemical composition is similar in all regions of the planet. Also note that all of us internally possess chemicals such as arsenic, lead, gold, copper, silver, mercury, tin, fluorine, chlorine, and uranium that in slightly larger amounts would prove fatal. With this information in mind, one can see the tremendous difficulties and challenges that confront research chemists in determining the effects other foreign natural and synthetic chemicals may have if introduced in minute quantities in our bodily systems.

The chemical industry dutifully and consistently examines the physical effects chemicals may have on humans regarding present and newly developed chemicals. Looking at public health trends, government reports, and environmental group queries, all branches of the chemical industry, including plastics, take the well-being and health of the public seriously.

Examples of the plastic industry's actions to address possible harm to the public occurred when excessive lead levels were found to be harmful, and the plastic industry moved quickly and purposefully to ban lead from plastic stabilizers; when vinyl chloride was discovered to be a possible carcinogen in the mid-70s, the plastic vinyl industry completely changed its manufacturing process to practically eliminate this chemical compound and install monitoring devices in all Polyvinyl Chloride (PVC) processing plants; and when bisphenol A and phthalates were suggested to be harmful, the industry funded and performed dozens of research projects to determine the effects of these chemicals on public health.

The plastic industry is tightly managed and regulated by several code bodies and governmental agencies to prevent harm to industry workers, the environment, and the public. In documented studies, the plastic industry has consistently proven to be much lower in reported illness and injury rates (IIR) than the overall manufacturing sector (in the case of the vinyl industry, IIR are over 80% less than the average).[4] Life Cycle Assessments (LCAs) are scientific studies that follow and evaluate the environmental impact of a product from cradle to grave. Plastics in most cases provide minimal environmental impact to the planet and, in most cases, have less of an impact when compared to competing materials. Several well-intentioned but uninformed environmental groups are advocating outlawing some plastics due to a perceived environmental safety threat. The results of LCAs and other scientific facts prove this concern to be without merit.

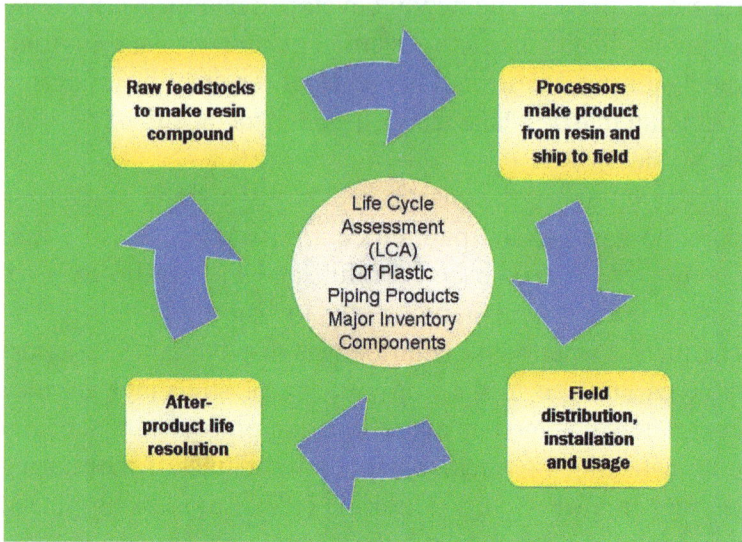

Figure 3 Life cycle assessment.

Not only are plastics environmentally sound, they are also cost effective. In a recent study it was found that to use competitive materials to replace plastics would cost the North American economy billions of dollars per year.[5] And with the economy still being in the doldrums, an anti-plastic movement would greatly increase building costs while exacerbating unemployment.

Man-made chemicals—plastics—have been in use for over a century. They have replaced many other chemical groups such as metals, glass, wood, clay, concrete, fur, hides, and ivory for four main reasons: they are more durable, easy and safe to install and maintain, environmentally sound, and cost-effective. Odds are that as new technologies emerge to solve the challenges of energy and pollution issues, plastic chemicals will be the material of choice for many of these future applications.

Notes:
1: Adapted from the definition of *The American Heritage Dictionary*.
2: From *The Elements*, 3rd edition by John Emsley
3: From *Elemental Composition of the Human Body* by Ed Uthman
4: From *U.S. Bureau of Labor Statistics of 2006*
5: From *" The Economics of Phasing Out PVC"* by Frank Ackerman and Rachel Massey of Global Development and Environmental Institute of Tufts University- Dec. 2003

3

THE CHINA STUDY

For Christmas I received a gift of a book that got me thinking: *The China Study*, by T. Colin Campbell and his son, Thomas Campbell. Author of over 350 research papers, the senior Campbell is Professor Emeritus at Cornell University in nutritional biochemistry and is in the forefront of nutrition research. *The China Study* that the title refers to is the most comprehensive study of health and nutrition ever undertaken. It comprises a survey of diseases and lifestyle factors in rural China and Taiwan begun in 1983 and continuing still. *The China Study* investigates the relationship between diet and the risk of developing disease, and looks at the connection between nutrition and heart disease, diabetes, and cancer. It also examines the source of nutritional confusion produced by powerful lobbies, government entities, and opportunistic scientists and groups.

Figure 1
Book co-author:
T. Colin Campbell.

The book was an interesting read. The conclusion, backed by listed scientific studies, is that people who ate the most animal-based foods and had a sedentary life-style got the most chronic disease, and that people who ate the most whole plant-based foods and had an physically-active life-style were the healthiest and tended to avoid chronic disease.

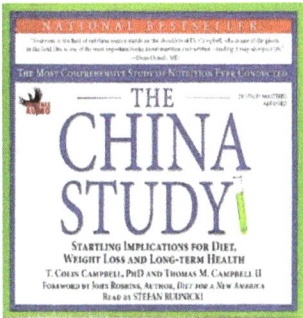

Figure 2 *The China Study.*

Something Dr. Campbell wrote focused my attention. On page two of his book he states that based on his experience, "Synthetic chemicals in your environment and in your food, as problematic as they may be, are not the main cause of cancer."

Giving that statement some weight is the fact that during the course of his career, Campbell was instrumental in the discovery of Dioxin and Aflatoxin, two of the most potent carcinogens ever found; he was also involved in research into the following alleged carcinogens: Aminotriazole (herbicide used on cranberry crops), Alar (apple spray), DDT, Nitrites, Red Dye Number 2, artificial sweeteners.

Campbell's opinion is that scientific conclusions from carcinogen studies that yield only marginal results in laboratory test animals are questionable, not definitive. Yet these questionable results make very big waves in the public arena. His claim is that the majority of cancers are caused instead by poor diets and passive lifestyles. He believes that to funnel research dollars into sometimes outlandish investigations trying to prove a carcinogenic link to a chemical is a waste of resources.

Campbell takes a humorous turn to illustrate his point, when on pages 44 through 46 he writes how absurd and flawed many of the studies of possible carcinogens are. He refers to a particular study of nitrites; a substance used in the processing of hot dogs, bacon, and preserved meat. Nitrites can form N-nitrososarcosines (NSAR), which may be a human carcinogen. The study involved the testing of rats in two groups. In the first, the rats were given a high dose of NSAR; in the other, a low dose (half the amount of the high dose). The result was that all the rats given the high dose died prematurely, but that "only" 35% of the low-dosage rats died, all of throat cancer.

A reasonable question here is how much NSAR were the rats fed? Campbell puts the dosages in layman's terms. Let's say you want to give someone throat cancer using NSAR compounds that exist in bologna. You would have to feed an individual 270,000 bologna sandwiches, with a pound of bologna per sandwich. Assuming the eater ate a sandwich at each meal, it would take approximately 25 years to digest the entire amount of the suspected carcinogenic substance, in this case NSAR. If your victim accomplishes this gluttonous feat, he will have had as much exposure to NSAR (adjusted for body weight) as the rats in the *low* dose group.

After reading Campbell's book, I considered the international furor against plastics, especially Polyvinyl Chloride (PVC). The fear that most of us feel toward cancer is capitalized on by special interest groups (SIGS). They discredit products for reasons that are often not clear or sound.

To me it doesn't make sense to mount a campaign against PVC (or any other plastic) as an alleged carcinogen. The spotlight should focus more fittingly on the origins of most cancers, poor lifestyle choices such as alcohol consumption, tobacco use, lack of exercise, and a diet high in dairy and animal fats. It also isn't convincing to condemn PVC resin plants, many located in Texas, as being producers of cancer causing chemicals, when the state with the highest incidence of cancer per 100,000 population is Maine (Texas isn't even in the top 30). [1]

Figure 3 Healthy fruit and vegetables help prevent disease.

I think the vinyl industry ought to invite Dr. T. Colin Campbell to a gathering and ask him to address the myths, lies, and half-truths perpetrated by SIGS against plastics. Perhaps his involvement could add additional credibility and prestige to the promotion of the industry's efforts, similar to that added by vinyl spokesperson Dr. Patrick Moore, a co-founder of Greenpeace.

Note[1] National Cancer Institute 2006 data

4

DEFINING BIOPLASTIC TERMINOLOGY

Bio is derived from bios, which in Greek means life. There are many common terms we are familiar with that use this prefix, such as biography, biology, biorhythm, and biosphere. In the plastics industry there is a host of bio labels whose use at times cause confusion, especially to consumers. The purpose of this article is to provide clarity to many of these bio-terms.

Biomass is a renewable energy source composed of recently living plant and animal organisms and can be used for the production of chemicals (plastics). Biomass plants such as switchgrass, hemp, corn, sugarcane, sorghum, eucalyptus, and oil palm are being studied and, in some cases, being converted into feedstocks for their own or other plastic resins. Although fossil fuels such as coal, gas, and petroleum have their origin in ancient biomass, they are excluded from the presently used biomass terminology because they contain carbon that has been out of the carbon cycle for eons.

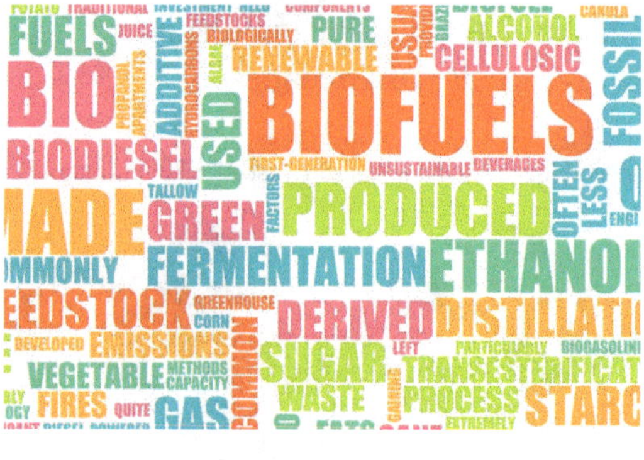

**Figure 1
Biomass sources
chart.**

Bio feedstock is a biomass raw material used in the man-ufacturing of a product. Almost all of the plastics manufac-tured today use fossil fuel feedstocks (natural gas, oil, and coal). However, the movement toward protecting and improving the environment is quickly moving the plastic industry to explore and embrace bio feedstocks for com-monly used plastics such as ethylene and propylene. An example of this present technology is the use of sugarcane in the production of ethylene, a basic feedstock for produc-ing polyethylene and polyvinyl chloride.

**Figure 2
Sugar cane feedstock
for plastics.**

Bioplastics and biopolymers are plastics that incorporate the majority of their feedstock as bio-based, non-fossil materials. These plastics are formed from three major types of biomass types: starch, sugar, and cellulose. Biopolymers for the most part are biodegradable, made from renewable sources, and modified chemically or with bacteria to pro-duce a finished product that has thermoplastic properties. Examples are: Polylactic acid (PLA) made from cane sugar or glucose and used in many packaging applications; and Cellulose Acetate (CA) and Cellulose Acetate Butyrate (CAB) made from such cellulosed plants as cotton and wood and used in tapes, packaging, and many other applica-tions. As noted before, other bioplastics use biopolymers to produce an organic compound which may be converted into a feedstock material for synthetic and mostly non-biodegradable plastic materials.

Bio-based contents of a plastic material have been quantified by using ASTM standard D6866. This standard measures the content percentage of carbon from renewable sources versus carbon from fossil carbon. Although bioplastics are expanding at double digit growth rates, they are still just a very small percentage of the overall 500 billion pound global plastic market (2010 estimate).

Biodegradable plastics are plastics that biodegrade from a process naturally occurring in the environment such as bacteria, fungi, and algae, and that are quantified by ASTM D6400 and EN 13432 standards. Quantifying degradation is important because all renewable and fossil fuel based plastics do degrade. However, many plastics degrade at such a slow rate as to be listed as non-biodegradable. Although in several cases a bioplastic is also biodegradable, there are bioplastics that do not degrade per ASTM and EN definitions. Conversely, some petrochemical-based plastics can be biodegradable.

The biodegradable plastics are mostly used in flexible disposable consumer products such as diapers, blister foils, and other packaging. According to the United Kingdom's NNFCC (National Non-Food Crops Centre), the growth of biodegradable plastic has increased by over 600% in the years 2000 to 2008 and could average a 17% growth rate through 2012. Yet, biodegradable plastics today are still a minute percentage of the entire market.

If trends continue, the next one or two decades will have plastic offerings available in renewable and sustainable bio feedstocks to produce two types of finished products: degradable and non-degradable. The idea is that disposal items like packaging, waste bags, diapers, and other items are allowed to degrade at a designed rate while durable items such as electronic casings, car interiors, plastic pipe,

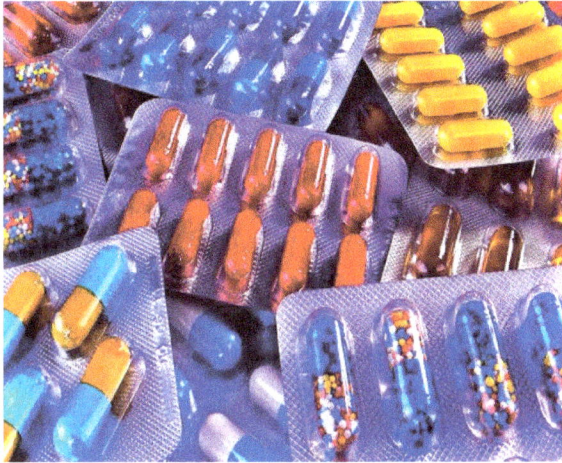

Figure 3 Plastic blister packaging.

window profiles, and other items are non-degradable and designed for maximum product life.

One caveat to consider, however, before jumping on the bio-based plastic bandwagon is for the scientific community to compare each present and replacement application material using life cycle assessment (LCA). It is quite possible that after examining complete bioproduct life cycles—including deforestation, raw materials, plant fertilizers and pesticides, transportation, and waste disposal—non-renewable plastics may be more environmentally sound.

Reprinted with permission of the IAPD; issue august/september 2012 – **the IAPD Magazine**

THE HISTORY OF NORTH AMERICAN
PLASTIC PIPING DISTRIBUTORS

Most industrial markets and products proliferate due to the vision and entrepreneurial passion of a manufacturing-oriented zealot. Henry Ford, Thomas Edison, Thomas Watson, Bill Gates, and Michael Dell are just a few examples that come to mind. However, it's rare in the annals of American Industry one can find the meteoric growth of an industrial market owing its success to a "third party" group of entrepreneurs—the plastic piping distributor (PPD).

The North American thermoplastic industrial piping systems (TIPS) market began in the mid 1950s with the introduction of vinyl piping products by a handful of manufacturers. Tube Turn Plastics, a joint venture of Tube Turns, Inc., (a steel-welded pipe fitting manufacturer) and Jackson

Figure 1 Distributor loading docks.

**Figure 2
Distributor yard with
pallets of plastic
piping.**

& Church (a plastic molding machine builder) in Louisville, Kentucky, became the first manufacturer to mold Schedule 80 fittings and valves. Later in 1964, Cabot Corporation bought Tube Turn's division and changed the entity's name to Cabot Piping Systems (CPS). With the financial backing of Cabot and the acquisition of Kraloy-Chemtrol, CPS soon became a dominant producer of TIPS products in the United States. During the late 1960s and early 70s, other thermoplastic pipe, fittings, and valve companies immerged as players in North America. A listing of most of these companies is shown in Table 1.

The major channel of distribution for these manufacturing companies was "metal houses" that predominately stocked, promoted, and sold standard and alloy steel piping products and a very small amount of TIPS. In actuality, the major reason these distributors inventoried and sold any TIPS products at all was due to the efforts of CPS and other company factory sales forces who had helped specify TIPS products.

In the late 1960s and early 1970s, a phenomena occurred—quite accidentally at first, and then with a more directed effort. It seems there were many entrepreneurial associates at CPS who saw an opportunity to get involved in distributing thermoplastic piping systems, mostly with core

CPS products in a geographical territory they were familiar with. A dozen individuals left Cabot in a 3-to-5 year span to spearhead the movement to increase market share of thermoplastic piping systems. A listing of these and other PPDs are shown in Table 2.

By the end of the 1970s, PPDs had approximately eighty-five percent of all North American TIPS sales. The estimated TIPS sales by 1980 were well over $150 million dollars with the CPS "incubated" distributors accounting for most of this total. The success of the PPDs, in most part, was due to the owners focusing on selling plastic piping systems only. They undertook many of the typical manufacturer's functions, including assisting in writing specifications, giving technical seminars, being available for job start-ups, and, in general, acting as the plastic "guru" in their geographical area.

Also, the newer industries were much more receptive to new piping materials. The semiconductor, electronic, photographic, metal finishing, waste and water treatment, water theme parks, aquariums, fish farms, and other high-tech and high-growth industries welcomed new cost-effective piping solutions. Last, an influx of innovative manufacturers (many foreign) had added a large variety of new products to complete a full distributor offering of thermoplastic pipe, fittings, valves, tanks, pumps, tubing, filtration, ducting, and other products in several different plastic piping materials.

The beginning of the twenty first century has unveiled a new look to the TIPS industry. There have been many consolidations both in manufacturing and distribution. There is no more Cabot Piping Systems (since acquired by Celanese Corporation in the 1970s and later by NIBCO INC, in 1981). And R & G Sloane is gone (acquired by George Fischer), as is Plastiline (acquired by Robintech, then Colonial Engineering, and last LASCO Fittings). Walworth and Jamesbury valves have closed shop while Carlon changed focus from TIPS products to mainly electrical conduit and mining piping products. Two rather innocuous companies, which began selling products in 1969, Spears Manufacturing and Asahi-American, have become very significant TIPS players in the last three decades. In addition, two large international plastic piping systems behemoths, George Fischer and Etex, have become a more dominant force in the U.S. TIPS market.

There has been even more consolidation in the plastic piping distributor arena. Two dominant PPDs—Harrington Industrial Plastics and Ryan Herco Products—have grown internally and from the acquisition of several independent PPDs. Today, there are about two dozen PPDs in the USA with over 100 branches and sales exceeding several hundred millions of dollars (over half of which are sales of Harrington and Herco). The PPD's market share of TIPS (especially vinyl products), however, has dwindled from the glory days of the 1970s and 80s to about fifty-five percent in 2008.

The PPDs, after doing much of the work in building the TIPS market, have had their market share diminished by the inroads of large do-it-yourself home retailers, plumbing supply company wholesalers, and other market channels that complement plastic piping products with non-plastic piping materials in order to offer a more complete package to buyers.

There are few phenomena in industrial manufactured goods that can significantly compare to the extraordinary success of growing the sales of an entire industry on the effort and foresight of about two dozen third party visionaries. It's doubtful that, without the presence and creativity of the plastic piping distributors, the TIPS industry would be as significant in the industrial market place as it is today.

Reprinted with permission of the IAPD; issue october/november 2012 —
the IAPD magazine

Table 1
TIPS PVF Manufacturers in North America Pre-1975

Manufacturers*	Products
Alpha Plastics	Pipe
Asahi-American	Pipe/Fittings/Valves
B.F. Goodrich	Pipe
Borg Warner	Pipe
Carlon	Pipe/Fittings
Celanese Piping Systems**	Pipe/Fittings/Valves
Colonial Engineering	Pipe/Fittings/Valves
E.I. DuPont	Pipe
George Fischer	Valves
Grinnell-Saunders	Valves
Harvel Plastics	Pipe
Hayward Plastics	Valves
Hills McCanna	Valves
Interpace	Valves
Jamesbury	Valves
Johns Manville	Pipe
Labline-Enfield	Acid Waste Drainage Systems
LASCO	Fittings
National Pipe & Tube	Pipe
Phillips-Driscopipe	Pipe
Plastiline	Fittings/Valves
Plastinetics	Fabricated Fittings
Plast-O-Matic Valves	Valves
Precision Polymers	Pipe
Robintech	Pipe
Ryerson	Pipe
R & G Sloane	Pipe/Fittings/Valves
Scepter Manufacturing	Pipe
Spears Manufacturing	Fittings
Vulcathene	Acid Waste Drainage Systems
Walworth Valves	Valves

Notes:
*Highlighted companies have been purchased by or merged into others, went out of business, or eliminated TIPS products from their offerings
**Celanese acquired CPS in 1974. NIBCO acquired the division from Celanese in 1981.

Table 2
Pre-1975 Plastic Piping Specialist Distributors

Distributor*	Location	Original Principals**
Aetna Plastics	Cleveland, OH	Paul Davis
Ayer Sales	Boston, MA	Bruce Ayer
Burt Processing	Hartford, CT	Bill Burt
Bushnell-Cicero	Chicago, IL	Ted Bushnell/Joe Cicero
*Chem-Pipe	Philadelphia, PA	Joe Hasiak
*Corro-Flo	Louisville, KY	Bill Vanegas
Corrosion Products	St. Louis, MO	Jon Schiller
Corr-Tech	Houston, TX	Jim Gottesman
*Don Drake & Assocs.	Dallas, TX	Don Drake/Mike Gladden/ Marion Whiteside
Eagle Supply & Plastics	Appleton, WI	Dave Ballin
Fabco Plastics	Toronto, Canada	M.K. Kehren
*Filter-Chem	Alhambra, CA	Will Ditmar
*Flonetics	Philadelphia, PA	Stow Shoemaker
*G.S. Comstock	Philadelphia, PA	George Comstock
*Gulf Wandes	Baton Rough, LA	Dudley Atkinson III
Harrington Industrial Plastics	Anaheim, CA	Marv Harrington, Cliff Springmier/Larry Collier/ Pete Youngdal/Bob Kenyon
M.L. Sheldon	New York, NY	M.L. Sheldon
*Malcom Black	New York, NY	Malcolm Black
*National Molded	Jersey City, NJ	Manneth Shear
*Norrell Plastics	Memphis, TN	Dudley Atkinson Jr/George Burroughs/Fred Beckendorf
Pena-Plas	Scranton, PA	Andy Bubser
Plastico	Memphis, TN	Herb & Alvin Notowich/ Larry Welch
*Plastic Piping Systems	Newark, NJ	Dennis Garber/David Chasis Kenneth Pollack/Ted Vagell
Plastic Supply & Fabrication	New Orleans, LA	Al Malone
Ryan Herco Products	Burbank, CA	Mike Ryan/Brian Bowman/ Terry O'Brian/Ed Glossup
Seeyle Plastics	Bloomington, MN	Richard Seeyle
Southern Industrial Supply	St. Petersburg, FL	G.W. Fine
Tenn-Plast	Memphis, TN	Ernie Sutherland
U.S. Plastics	Lima, OH	Stanley Tam

Notes:
*Highlighted companies have been purchased by or merged into others, went out of business, or eliminated TIPS products from their offerings
**Celanese acquired CPS in 1974. NIBCO acquired the division from Celanese in 1981.

INTERNET ... CURSE OR BLESSING FOR PLASTICS?

The speed and volume of digital informational communication is astounding. In the past several years, we adopted Wikipedia instead of encyclopedias; Google instead of libraries; MapQuest instead of maps; spell-checkers instead of dictionaries; iPods instead of record players; online shopping instead of malls; e-mail instead of letters; and texting instead of talking. As Bob Dylan sang, "The times they are a-changin'."

Figure 1 Frustrated Internet user.

Today more data is available at the touch of one's computer keyboard than ever existed in all the combined world libraries. OK, so we have access to more information than we can possibly use. Does this make us wise? Does this allow us to make better and more informed decisions? Yes and no. Yes, if we can verify the authenticity and accuracy of that information. No, if we're working from faulty data.

The plastics industry—now more than a century old—was founded on advances through science. Organic chemists and researchers work diligently and methodically in exploring and uncovering the secrets of hydrocarbons. The creation of new, useful plastic compounds is based on fact-finding and experimentation. Yet, throughout the past several decades, the industry has been subject to critics decrying that advancement, based on unverified research and inaccurate conclusions. Today, the primary tool used to spread misinformation is the same one used to gather so much valid information—the Internet.

In the past several years, Internet attacks on the plastics industry have ranged from random and easily discounted statements to more concentrated disinformation campaigns from well-funded, coordinated groups. Such ludicrous statements have included claims that vinyl shower curtains give off dangerous gasses; plastic water bottles can cause cancer; and heavy metals can be leached out of PVC pipe.

The U.S. Food and Drug Administration has investigated and dismissed the majority of these types of claims. Globally, many scientific communities have tested and approved the use of most plastic products as safe and effective. Extensive documented research—performed not only by the chemical and plastics industries, but also independently by third parties—shows the error of many of these anti-plastic statements. Yet, the spread of misinformation continues aided by new technologies.

Faced with a nebulous cloud of dubious charges, how can the industry best respond? It needs clarity, certainty, and specificity—and it must be at least as adept in making use of technologies and social networking as are those spreading bad information. Here are four strategies that can return sense and credibility to the picture, no matter whether one views that picture on a PC or IPhone.

Figure 2 Internet marketing confusion

Make Motives and History Clear

How many consumers realize the long history of safety and careful investigation of issues the plastic industry supports? The industry has a historical reputation for researching and reporting on any and all claims of personal and environmental harm possibly attributable to plastics. Past experiences have shown that if research proves any of the claims to be true, or even having a hint of being true, the plastic industry will aggressively terminate or amend processes and/or materials to protect the health and well-being of the public.

The industry has never demonstrated an attitude of apathy or of "putting its head in the sand" when confronted by potential health issues. The plastic industry's actions and motives regarding health concerns needs to be explained, documented, and promoted to the public to combat negativity.

Aggressively Promote the Benefits of What is Produced

Another strategy is to share with the public the many features and benefits plastics bring to their world and how plastics save energy, water, and even lives. The plastic industry has a great message; what is needed is an investment in getting these messages into the public's eyes and ears.

Counter Charges with Specific Facts and Data

The plastic industry can play a role in quelling myths and inaccuracies by directing the public to established web sites that scientifically and directly address anti-plastic propaganda. Best of all, most of these are easily accessed online. Industry group web sites such as the American Chemistry Council's *plasticsmythbuster.org* is already having a significant effect on changing the conversation. And popular independently secular sites that address Internet and e-mail rumormongering such as *snopes.com* and *breakthechain.org* are already looked to by many to get the facts straight.

Just as industry support groups are doing, individuals can address inaccuracies as they are encountered in their work and lives. This doesn't mean getting into "wars" where accusations are traded for days with random Internet commenters (these usually result in a waste of time and effort). In fact, some people deliberately post provocative e-mails and comments in order to incite arguments—it's their idea of fun. An effective policy is to simply respond to an e-mail, comment, or article with a reference to a scientific citation or a reputable website that has the facts. Another tactic is to forward such remarks to industry associations such as the American Chemical (ACC), Vinyl Institute (VI), Plastic Pipe Institute (PPI), Plastic Pipe and Fittings Association (PPFA), International Association of Plastics Distribution (IAPD), or other comparable organizations. It would also be prudent to become actively involved with professional organizations to work on ways to solve this challenge.

Use the Same Tools as Those of the Adversaries

As individuals and working with professional associations, the public can become skilled at using new technologies and social media. Read and comment on blogs, subscribe to Twitter or Linkedin, start a web site that gathers the best scientific information. How about a YouTube video praising the environmental advantages of plastics or a Facebook page called "Plastic is Good for Me?" Be creative and get the message out to end-users and the engineering community.

The plastics industry has a positive and compelling story to tell—whatever the media. What is needed now is for the industry to address the challenge and to commit the resources so it can robustly take advantage of the digital information age, making the Internet a blessing, not a curse, for plastics.

Reprinted with permission from IAPD; issue june/july 2009 – **the IAPD magazine**

7

PEARLS OF PLASTIC

One of my professors in graduate school once quipped, "Statistics are like bikini bathing suits–what they show is revealing, but what they hide is vital." This technique of presenting statistics in a biased way to prove an argument is evident in Bryan Walsh's article, *Perils of Plastic,* published on April 1, 2010 by *Time Magazine.*

The article begins with Walsh celebrating how much the quality of air and water in the United States has improved over the last four decades. But then without using research or facts, he goes so far as to suggest that "Americans *may be* sickening" [Italics mine]. Loaded, unsubstantiated phrases such as "*may be sickening*" play to the readers' emotions, but do not contribute to furthering the author's arguments in any logical, scientific way. Peppered throughout Walsh's article are many of these phrases: "*may disrupt,*" "*could have,*" "*may have,*" "*may mess,*" "*a possible risk,*" "*which might,*" etc. Yet nowhere in the article does he substantiate his hypothesis that as the land and waters have been healing since 1970, Americans are seeing setbacks in health due to the use of plastic.

The truth is that Americans have been extending their life expectancy, while the usage of plastics has grown. A little more than a hundred years ago, when the age of synthetically made plastics began, the life expectancy of Americans was 50.1 years. Today, in 2010, our life expectancy is 78.3 years, a whopping 56.3% improvement. Although the United States ranks 38th in world life expectancy, several other developed countries with relatively corresponding per capita plastic consumption and production facilities as the United States are in the top tier of countries with the highest life expectancy, including Japan at 82.6 years, France at 80.7, Canada at 80.7 years, Italy at 80.5, and Germany at 79.4 years[1].

Nowhere in the scientific literature is there proof that plastics and other industrial chemicals cause lower life expectancy, either for Americans or for citizens of the other nations that are also large manufacturers and users of plastics.

With the advent and use of Life Cycle Assessment (LCA), the recognized and accepted scientific method of quantitatively assessing environmental impact of products from cradle to grave, one might effectively argue that plastics have helped extend lifespan of humans and other species. Consider the materials that plastic has replaced in such fields as building, packaging, furniture, clothing, transportation, and industry. What materials did plastics replace? Glass, wood, concrete, vitrified clay, ceramics, ivory, animal hides, and furs, as well as metals such as aluminum, brass, copper, cast iron, lead, steel, and tin, to name a few. In almost all LCA comparisons of plastics to the materials they replaced, plastics make significantly less of an environmental footprint on the planet.

But according to Walsh, scientists can't be trusted. He stated in his article that government regulators (FDA) are the final judges of approving drugs and pesticides for public use. However, when it comes to chemicals (plastics), he advocates that the public, not scientists, should interpret the body of scientific data. In other words, he is saying the public is better equipped to make informed technical decisions about science than those conducting the tests. His suggestion brings sharply to mind many erroneous conjectures and subsequently dispelled myths, such as plastic shower curtains emitting toxic fumes; microwaving foods in plastic containers releasing cancer-causing agents into the food; and dioxins being released by freezing water in plastic bottles.

If Walsh really has an altruistic intent to improve the health of Americans, I would submit that he spend time reading and interpreting the internationally acclaimed best seller, *The China Study*.

In that book, Doctors T. Colin Campbell and his son document the most comprehensive study of health and nutrition ever undertaken, a survey of diseases and lifestyle factors throughout China, begun in 1983 and continuing today. The book investigates the relationship between diet and the risk of developing illnesses such as heart disease, diabetes, cancer, and other chronic diseases. The book's conclusion, backed by dozens of scientific studies, is that people who ate

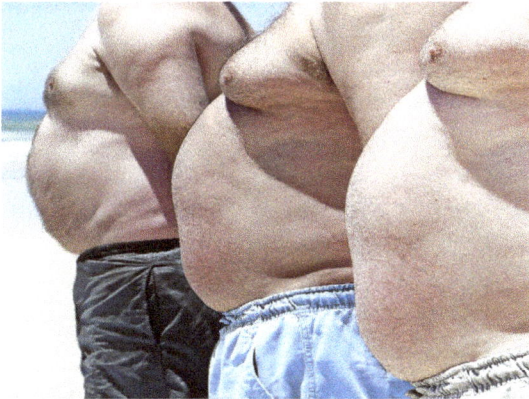

Figure 1 Poor diet and life style shortens life-span in many developed countries.

the most animal-based foods and had a sedentary life style got the most chronic disease, and that people who ate the most whole plant-based foods and minimally exercised were the healthiest and tended to avoid chronic disease.

But Campbell goes even further in stating that scientific conclusions from carcinogen studies yielding only marginal results in laboratory test animals are questionable, not definitive. He believes that to funnel research dollars into sometimes outlandish investigations trying to prove a carcinogenic link to a chemical is a waste of resources. Elsewhere in his book he states that, "Synthetic chemicals in your environment and in your food, as problematic as they may be, are not the main cause of cancer." Again, he concludes that the majority of cancers are not caused by plastics or other chemicals, but by poor diets and passive lifestyles.

Mr. Walsh, for your information, plastic products as a rule are produced using less energy, are more durable and easier to install, weigh less (which reduces the energy required to transport), and, oh yes, are extremely cost-effective.

With sustainability, greenness, and job formation being the hot topics in today's news media and political arena, plastics should be looked at in a new, more informed light, not vilified.

Notes[1]: Data from the U.S. National Center for Health Statistics

PLASTIC PIPING SYSTEMS AND THE INTERNET

As more and more people use the Internet instead of books to search for answers and/or to gather useful information, there is a definite need to seek and identify the most reliable, current and user-friendly websites. This is especially true for the plastic piping industry, where many rumors and much anti-plastic, unscientific rhetoric abound. To simplify the search, here is a listing of several sites that are very useful. Two recently created sites in particular are now available and contain comprehensive information to help engineers, installers, end-user, students, building code officials and other audiences searching for information on plastic piping systems.

The plastic piping association websites that are chock-full of useful information are:

International Association of Plastic Distribution (IAPD)

This site contains dozens of articles on plastics in general as well as plastic piping systems in particular from their signature and highly recognized publication: *IAPD Magazine*. In addition, there is a listing of educational tools that offer basic information on dozens of various plastics for piping as well as sheet, rod, and tubing. www.iapd.org

Plastic Pipe and Fittings Association (PPFA)

This site has technical and installation workbooks, videos, manuals, and white papers (all of which can be downloaded) on the most commonly used plastic piping system materials for residential, commercial and industrial construction—this includes Acrylonitrile Butadiene Styrene (ABS), Chlorinated Polyvinyl Chloride (CPVC), Polyethylene (PE), Cross-linked Polyethylene (PEX), Polypropylene (PP) and Polyvinylidene Fluoride (PVDF) piping systems. www.ppfahome.org

Plastic Pipe Institute (PPI)

Everything you want to know about PE piping is on this site. In addition, there are several excellent technical articles regarding most plastic piping materials. One of the most frequently used technical reports shown on the site is TR-19/2007, *Chemical Resistance of Thermoplastic Piping Materials*. TR-19 lists the chemical resistance capability of ten plastics with hundreds of chemical agents. www.plasticpipe.org

Uni-Bell PVC Pipe Association

If you are designing or installing PVC underground piping, this is the site to reference. It lists over two dozen publications and videos concerning buried PVC piping systems. Uni-Bell also lists on the site for purchase the industry's "Bible" of underground PVC piping usage—*The Handbook of PVC Pipe Design and Construction* (co-published by Uni-Bell and Industrial Press, Inc. www.uni-bell.org

Two relatively new sites that are highly recommended as sources for gathering information on plastic piping are: www.sustainablepipingsystems.com and www.opus.mcerf.org.

Sustainable Piping Systems

This site is a combination website and blog created and maintained by Chasis Consulting, Inc., to offer one source of information on plastic piping systems. Included on the site are over 40 articles, links to almost 100 relevant associations which have some connection to plastic pipe, listing and direct links to over 100 product manufacturers, a listing and linking of educational tools, useful engineering information, an on-site video library, and a section for comments and questions. This website/blog is dynamic in nature and will be refreshed on a monthly basis as new information becomes available. www.sustainablepipingsystems.com

OPUS

The Mechanical Contractors Association of American (MCAA) through its Mechanical Contracting Education Research Foundation (MCERF) has created a site that includes plastic piping as well as other piping materials. The site lists 17 different piping materials (8 are plastic), 24 different system applications, and 17 pipe joining methods. The beauty of the site is that an inquirer can research a particular piping material and/or application and not only get a summary of the requested material and application, but also a list of which piping material is recommended for which system application. This easy-to-use site is an excellent guide to those who are just learning about piping materials as well as experienced engineers and installers who are open and receptive to new piping materials and applications. The only defect of the site is that it doesn't list most of the piping product manufacturers, but rather lists only those product manufacturers which belong to the MCAA. But using the sustainable piping systems website (which lists over 100 product manufacturers) in conjunction with this site should solve this imperfection. www.opus.mcerf.org

There are many other sources for gathering information about plastic piping systems such as searching the web for particular manufacturing websites, several of which contain useful technical information. However, for those who want a convenient and comprehensive source of readily accessible information on plastic piping systems, these six listed sites should fit the bill.

Reprinted with permission of the IAPD; issue august/september 2010 – **the IAPD magazine.**

PLASTICS AND PETA

You wouldn't think that plastics and PETA (People for the Ethical Treatment of Animals and pronounced like the Greek pocket bread) have a lot in common; but, they most certainly do.

PETA is a non-profit advocacy group with a few million members and an annual budget over $25,000,000. One of the group's basic doctrines is not to use animal pelts or other body parts for human comfort or entertainment. So where does plastics fit into the picture?

One of the first man-made plastics was cellulous acetate, whose prime purpose for discovery was to replace ivory in billiard balls. And it did! Ivory is a high impact material made from the tusks or teeth of elephants, hippopotami, walruses, and narwhals. It was also used for piano keys, bag pipe flutes, buttons, sculptures, combs, jewelry, and other

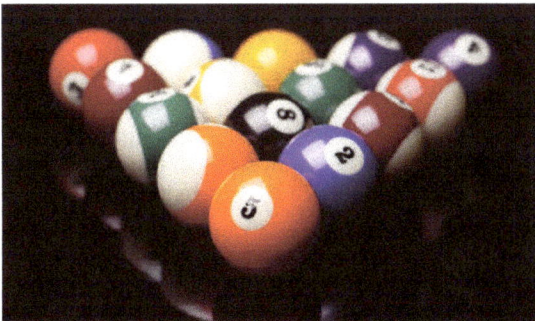

Figure 1 Plastics replaced ivory in the manufacture of billiard balls.

items. Plastic's growth has greatly reduced the need and demand for ivory, thereby saving the lives of thousands of the planet's largest mammals.

Another animal life-saving group of plastics is the flexible and resilient films and sheets of vinyl, nylon, and other polymers. Many times these materials are used as replacement for the hides of such animals as cattle, hogs, deer, lizards, alligators, and other animals in the manufacture of upholstery, shoes, belts, hand bags, luggage, clothing, and football helmets, to name a few. These synthetics are extremely cost effective; in many cases, it may take a professional to tell the difference between the animal leathers and the plastic look-a-likes.

Another group of plastics, mostly acrylics, are used to make faux fur. This substitute looks and feels like exotic animal fur and can outwear the natural covering of our furry animal friends. If there is a problem with these fake furs, it is that at first glance it is difficult to tell from the real thing! This feature could lead to an over-zealous protester who may mistakenly throw paint on your beautiful acrylic wrap or faux bear rug. So, when you purchase the next stuffed animal for a loved one, check out the cuddly fluffy exterior—yeah, it's a fur substitute: plastic!

Figure 2
It's tough to tell the difference
between real fur
and acrylic fiber.

PETA doesn't profess to be as spiritually attuned group as the Jainists, a religion which began in the 9th century B.C. in India. Jainism believes that any living thing possesses a soul worthy of respect. Therefore, it wouldn't be much of a stretch for PETA to come to the defense of the innocent silkworm. To make silk, the cocoon of the mulberry worm is used by destroying the pupae before the adult moth emerges. This is accomplished by boiling the pupae in water or piercing it with a needle to simplify unraveling the cocoon in one continuous silk thread.

Again, plastics to the rescue! The discovery, decades ago, of nylon and rayon as a replacement for silk has been successful in sparing the deaths of millions of moths-to-be. Not only has silk been largely replaced by plastics but the replacement materials are more durable, cost less, and have an inexhaustible source of supply.

To learn more about PETA and the dozens of companies that use synthetic materials in their manufacturing as a replacement for animal parts, go to PETA's website http://peta.org. To learn more about plastics and their use in our environment, click on the American Chemistry Council website http://americanchemistry.com

In the world today, plastics seem to be in targeted by several environmental activist groups who seem to ignore the many benefits plastic bring to our planet. What better rewards can any material group, other than plastic products, offer when it comes to being responsible for saving the past, present, and future lives of millions of earth's creatures?

Printed with permission from the IAPD; issue october/november 2008 – **the IAPD magazine**

10

POTABLE WATER FOR A HONDURAN VILLAGE

Background

With over half of its population living below the poverty line, Honduras is economically one of the poorest countries in the western hemisphere. Many small Honduran municipalities and villages do not have even the basic resources to provide safe, potable water systems for their citizens. For over 12 years, the people of one such village, Colinas de Suiza, spent a third to half of their income per family to purchase three 55-gallon barrels of water daily, delivered by entrepreneurs. Not only was the system expensive, but also the handling contaminated the delivered water.

The Humanitarian Engineering Program of the Colorado School of Mines (CSM) became involved in 2004 when it began designing a water distribution plan for Colinas de Suiza; it took on the project of assisting in the construction of a potable water system for the 1350 families (8,000 people) living there. Colinas de Suiza was established by the Honduran government to house refugees from the devastation of Hurricane Mitch in 1998. It is located in the hills of the Sula valley, within the municipality of Villanueva.

Figure 1
Villagers digging ditch for PVC piping.

The goal of the CSM project was to replace the existing water delivery system by pumping water from a nearby aquifer (400-foot-deep well) into a 250,000 gallon capacity storage tank. The water would then be distributed by gravity to the villagers' homes. When completed, the project will

Figure 2
Villagers covering installed PVC piping.

reduce the cost of water to 1/50th of that associated with the truck delivery, eliminate contamination, and reduce energy consumption by 95%.

Involved Institutions

In addition to CSM's efforts, several other institutions contributed by donating money and in-kind goods:

- CEPUDO (part of Food for the Poor in Honduras)
- Food for the Poor
- Mondialogo Engineering Award (collaboration between Daimler Chrysler and UNIESCO)
- Plastic Pipe and Fittings Association (PPFA)
- Universidad Autonoma de Honduras Valle Sula (UNAH-VS, a private Honduran engineering university
- Universidad Tecnologica Centroamericana (a Honduras engineering university)
- William and Flora Hewlett Foundation

By far the largest goods providers for the project were members of the PPFA. The project engineers specified Polyvinyl Chloride (PVC) as the piping material to use for the potable water system for several reasons: very durable, easy and safe to use, environmentally sound, and cost-effec-

**Figure 3
250,000-gallon water tank.**

tive. Plus, PVC has been successfully used to handle drinking water for over seven decades. PPFA member companies provided over 72 tons (45 km) of PVC materials including: 146,000 feet of piping, 6000 fittings, 1825 valves, 182 gallons of primer and solvent cement, and, most important, the services of an experienced installation supervisor. The market value for the donated goods and services exceeded $150,000. The PPFA member companies who participated in the 2007 shipment of goods were:

- George Fischer Sloane
- Hayward Industrial Products
- IPEX
- IPS Corporation
- J-M Eagle
- LASCO Fittings
- Mueller Industries
- NIBCO
- PipeLife Jet Stream
- Shintech
- Silver-Line Plastics

**Figure 4 Washing clothes from a
potable water tap.**

Project Challenges and Status

Finally, in early 2008, the Honduras project installation started in earnest, with labor provided by village volunteers.

Some on-site material thefts, the sudden illness and death of the on-site municipal Director of Water and Sanitation, three different mayors of the municipality of Villanueva, improper installation of the 6-inch-diameter PVC water pump line, and the occasional lack of local finances caused setbacks to the project. Fortunately, all these challenges were met and overcome. One item worthy of mention is that the people of Colinas de Suiza have not only contributed their labor, but also provided funds for the construction of the water storage tank and pumping system. Seventy-five percent of the village families each contributed an average of $100 to the project. This represents about 13 days of wages for the average laborer in Honduras.

**Figure 5
Quick shower with installed clean
potable water.**

The good news is that potable water is now being delivered from the storage tank to over 90% of the homes of Colinas de Suiza. Within a few months, all the families will have a running-water tap on their premises. From project conception to piped delivery of water to each home has taken almost 7 years. If you ask any of the 8,000 villagers, you will find that the wait was certainly worth all the hard work and challenges.

Reprinted with permission from the IAPD; issue december 2011/january 2012 – **the IAPD magazine**

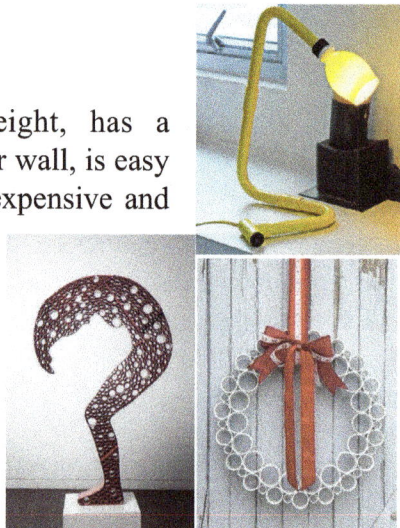

PVC PIPE USED FOR NON-PIPING PRODUCTS

Millions of miles of Polyvinyl Chloride (PVC) pipe have been installed on the planet, mostly for water mains, sewage, drainage, irrigation, swimming pools, and chemically aggressive applications. Why PVC? It's durable, easy and safe to install, environmentally sound, and cost-effective. Remember too, that all PVC products are completely recyclable. The stellar qualities of PVC pipe have inspired not only specifying engineers, but also designers and artists around the globe to use the piping products in numerous creative ways. We see PVC pipe put to clever decorative use, in functional and easy-care furniture, in recreational toys and games, for storage solutions, and in many other artistic and inventive applications.

Decorative Products

PVC pipe is lightweight, has a smooth interior and exterior wall, is easy to fabricate with rather inexpensive and simple tools, is available in a multitude of colors, and is not as ex-pensive as most other materials.

Figure 1 Desk lamp, pipe sculpture, and decorative wall hanging.

These characteristics make PVC a fascinating medium for designers and artists to experiment with in the creation of long-lasting decorative products.

Furniture

Chairs, tables, benches, swing sets, and cots, to name a few, can all be built using PVC piping. Just think of being able to lift furniture without incurring a hernia; or leaving patio chairs and table outdoors without worrying about rust or corrosion; and little-to-no maintenance is ever required. Now, that's what I call user-friendly.

Figure 2 Outdoor swing, end table,and table legs.

Recreational Games

Gaming products made from PVC pipe are available from commercial manufacturers, as well as designed and built by a DIY (do-it-yourself) public. If you can envision a recreational game product, the odds are good that PVC piping may have a place in its design. Musical instruments, tents, toys and games of all kinds, and even boats can be created using PVC pipe.

Figure 3 Outdoor towel rack.

**Figure 4
Bolo toss game, and
soccer goal.**

Storage Solutions

Wine racks, mailing tubes, shelving, containers for smaller diameter pipe, organizers, and many more PVC pipe storage solutions are readily available in stores. PVC piping is easily found at affordable prices in the plumbing aisles of large DIY retailers as well as smaller wholesale plumbing houses. If you have a storage idea, you can build it from PVC pipe. All you need to be creative is a drawing, pipe, fittings, cementing paraphernalia, measuring tape, and a saw.

**Figure 5 Exercise ball shelving, mail storage,
and clock embedded in plastic fitting.**

Miscellaneous Products

Imagination is a trait that all of us have. And with an idea and the versatility of PVC piping, there is hardly any limitation to what you can create and build, as you can see from this article.

Figure 6 Bench with sculpture, life-guard station, and chicken-feed trough.

So the next time you take a gander at PVC piping, think beyond its major use in effectively transporting fluids—think about the multitude of non-piping applications PVC can take on.

Reprinted with permission of the IAPD; issue april/may 2013 – **the IAPD magazine**

12

TRANSFORMATION FROM A RADICAL TO A
RATIONAL ENVIRONMENTALIST

The book *Confessions of a Greenpeace Dropout: The Making of a Sensible Environmentalist* by Patrick Moore, acclaimed author and co-founder of Greenpeace, was recently published; it presents in layman's terms what can go terribly wrong in an organization initially created to be a positive force in improving the environment and then morphing into an organization where fear, misinformation, and hidden agendas actually cause harm to the planet and its inhabitants.

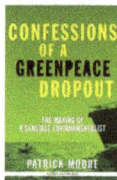

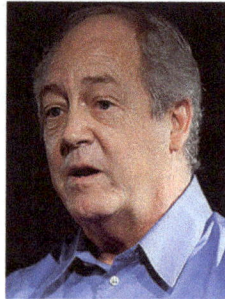

Figure 1
Author Patrick Moore
and hisbook.

In 1970 it was first known as *The Don't Make a Wave Committee,* a small group of radical activists whose meetings were conducted at a Unitarian Church in Vancouver, British Columbia. Shortly thereafter the committee changed its name to *Greenpeace*—and an environmental movement was born. The name Greenpeace was adopted by the co-founders in recognition of its first major campaign of eliminating the prospect of a nuclear war by banning nuclear weapons and testing so that a "green peace" could be realized.

In the book's earlier chapters, there are harrowing and intrepid tales of the under-funded and heroic battles the author and his cohorts waged against world powers in eliminating above ground nuclear testing, preventing and eliminating the killing of whales, stopping the inhumane slaughter of seal pups, and greatly reducing toxic discharges into the air and water. Greenpeace deserves much credit for these very successful campaigns to improve the earth's well-being.

In the mid-1980s, as Greenpeace grew into a global organization with branches throughout the world and hundreds of millions in revenues, it began heading in a direction that Dr. Moore (Ph.D. in Ecology) felt was very counterproductive. He concluded that Greenpeace's hierarchy had decided to pursue an agenda of choosing militancy over diplomacy, propaganda over science, dogma over compromise, and "nature" over human-beings. Try as he did, he determined that he could not change the minds and methods of Greenpeace's leadership.

After 15 eventful years of being a Greenpeace cofounder and leader, Patrick Moore left and began advocating for environmental causes he felt he had a firm grasp of the background science and could use a common sense approach of consensus-building and problem solving. He formed a consulting firm (named *Greenspirit Strategies Ltd.* in 1991) and today remains a force in using science and reasoning to solve some of the planet's most vexing problems. Listed below is a summation of his unshakeable and researched

beliefs stated in his book:
- Grow more trees and use more wood
- Use hydroelectric energy wherever possible
- Nuclear energy is essential for future growth
- Geothermal energy is an important and significant renewable energy source
- Develop technologies that use less or no fossil fuels to operate
- Genetic science improves the well-being of people and the environment
- No killing or capturing of whales or dolphins
- Aquaculture (fish farming) is a boon to mankind
- There is no undue cause for alarm concerning climate change
- Poverty is the world's worst environmental problem
- Most campaigns against useful chemicals are based on fear and misinformation

In regard to the last belief, Dr. Moore vehemently and continuously rails against Greenpeace's vicious campaign against chlorine, which they call the "devil's element," PVC which they refer to as "the poison plastic," and other chemical compounds.

He has pointed out to many audiences that chlorine is:

- The 11th most abundant element in the earth's crust
- Used in over 75% of all pharmaceuticals
- Mixed with water to produce hydrochloric acid which aids in digesting human food
- Added to drinking water, making it the biggest advance in public health—ever

His defenses against anti-PVC lobbyists include the following points:

- PVC's fire resistance makes it the preferred material for electrical insulation and conduit.
- PVC pipe leaks and corrodes less than concrete or metal piping systems and costs less.
- PVC manufacturing facilities produce less than one-half of 1 percent of all dioxins in the environment.
- PVC does not even rate as one of the top 10 dioxin producers in the world.
- PVC is used in hospitals since it is impervious to germs and easy to disinfect.
- PVC resin and compounds have never been proven toxic by any credible scientific organization.

The Confessions of a Greenpeace Dropout has a lot of convincing arguments in its 380 pages. When one finishes Moore's book, it is clear that the recurring theme throughout his thought-provoking writings is looking at environmental challenges logically and scientifically balancing the needs and values of humans and their environment. In this way, any radical environmentalist can be transformed to a rational one.

13

VINYL AND THE PLANET OF THE APES

There is a conspiracy afoot on this planet initiated and maintained by what I would call the Anti-Progressive Extremists Society or APES for short. This group deals in fear and propaganda as tools to attack a chosen subject. APES members are also perennial pessimists. They would be the ones who would find fault with Halle Berry's looks, Ernie Els' golf swing, Mozart's sonatas, and Michelangelo's sculptures. They would be a laughable and pitied lot if their attack campaigns were ignored. But alas, many of their unproven, unscientific, and unscrupulous accusations are peddled by the Western media who, looking for sensationalism at the expense of facts, give them a bully pulpit. One of the smear campaigns critics have undertaken in recent years is to attack plastics in general and vinyls [Polyvinyl Chloride (PVC) and Chlorinated Polyvinyl Chloride (CPVC)] in particular.

Vinyl is one of the most ubiquitous materials in use today. It is used in packaging, clothing, furniture, medical supplies, automobiles, and dozens of other industries, but its major use (65% of total PVC poundage) is in the construction industry. This industry's products include pipe, ducting, siding, fencing, decking, wire insulation, conduit, window profiles, and valves. The construction market has embraced vinyl. In many cases, it has made vinyl one of their preferred materials because of its proven benefits of durability, being easy and safe to use, environmental friendliness, and cost-effectiveness. Now, you would think that a material that uses less energy to produce, has less emissions of CO_2 in its manufacturing process , doesn't have to be replaced for decades, and has one of the lowest installed cost of any building material would be lauded by critics; but it is not the case. Instead, they seem to have an unbridled desire to see vinyl banned from use everywhere. Let me share with you their attack issues and the facts.

Chlorine is bad. Of course this statement ignores that chlorine is one of the most plentiful elements in the world, a major ingredient in salt, bleach, and many pharmaceuticals; arguably, it has saved more lives in its use of disinfecting water for drinking purposes than any other world health initiative—*ever.*

Vinyl creates dioxins. Dioxins are compounds which are suspected of being human carcinogens. The anti-PVC advocates claim that vinyls pollute the environment with dioxins. In reality, the largest contributors of dioxins (USA-EPA findings) are forest fires, wood burning fireplaces, coal-fired utilities, metal smelting, diesel trucks, sewage sludge land fills, and burning of trash. Some studies show vinyls produce 13 grams of dioxin a year (that's less than a half ounce). Another irrefutable fact is that dioxin levels in the USA have *decreased 90%* in the past 30 years while vinyl production has *increased 300%* during that same time period.

When vinyl burns, hydrogen chloride gas is emitted. This is true. And, when wood burns carbon monoxide is emitted. Are the extremists saying that one toxic gas is better than the other? Let's get real. With any residential or commercial structure you name, there are much more harmful fumes given off by other construction materials than by vinyls. Why? Because vinyl construction materials account for less than 5% by weight of most buildings.

Vinyl Chloride Monomer (VCM) could be harmful to humans and vinyl manufacturers are the prime source of emitting this gas. Forty years ago, there were reported incidents that excessive VCM, an intermediate material in the vinyl production chain, was excessively present in a few manufacturing facilities. The vinyl industry in cooperation with OSHA and the EPA completely re-engineered vinyl production facilities which have resulted in reducing VCM emissions by 99%.

Phthalates pose a threat to human health. Phthalate is a resin additive called a "plasticizer" which causes PVC material to be soft and flexible for such products as shower curtains and flexible tubing. There are *no plasticizers* in rigid PVC piping, so phthalates are not an issue. However, these public undocumented assertions harm all PVC products. There have been years of scientific studies in the United States and Europe concerning phthalates with no substantial data whatsoever showing they are a danger to human well-being.

Unfortunately science and facts don't seem to sell in our skewed media culture. But in spite of these factors, the vinyl industry is pursuing something called *Life Cycle Assessment* (LCA). LCA is a process which analyzes a product from birth to death. This process results in determining the energy, economic, and environmental impact of the entire product cycle in a quantitative manner. Preliminary and completed results of these studies show vinyl building products are the equal of or more favorable than other comparable materials.

Thomas Jefferson got it right when he stated that "Proselytizers dread the advance of science as witches do the approach of daylight." Vinyls by any objective measurement have been a phenomenal boon to our world. Let us all be aware of the misinformation and dogma of a cadre of well-funded and misguided activists so they don't create for us a planet of APES.

Reprinted with permission of the IAPD; issue august/september 2008 –
the IAPD magazine

PART 2

General Plastic Piping

GENERAL PLASTIC PIPING

The articles in this section demonstrate why plastic has become one of the most specified and installed piping systems on the planet. They also discuss what new technologies have emerged that benefit from the use of plastics:

14. *Advantages of Flexible Groove-Coupled Plastic Piping Systems* presents compelling reasons for using mechanically flexible groove-coupled systems in joining plastic piping.

15. *The Evolution of Pipe* documents the development of piping systems, from the earliest man-made systems of moving fluids to the present day preference for plastic piping in many applications.

16. *Fairy Tale or Reality* concludes that when you consider all the facts, plastic piping is one of the greenness piping materials available.

17. *How to Select the Proper Thermoplastic Industrial Piping Material* lists the basic methodology for choosing the proper thermoplastic piping material for a particular industrial application.

18. *The Ideal Pipe* explains why plastic, with all its features and benefits, should be at the top of the list if a group of engineers were going to design the perfect piping system.

19. *Joining Thermoplastic Industrial Piping Systems* provides a brief summary and description of the most common methods of joining various industrial thermoplastic piping systems.

20. *Join The Revolution...Think Plastics* offers several reasons why there is a revolution of metal-piping-focused engineers turning to plastic piping systems to solving project challenges.

21. *Life Cycle Assessment and Plastic Piping* explains what life cycle assessment is and how important this scientific tool should be used to evaluate all piping materials.

22. *Light Bulbs and Pipe* compares the environmental effects of compact fluorescent light bulbs to PVC piping systems.

23. *Plastic Piping in Industrial Applications* describes the reasons plastics have made inroads in industrial piping systems, and gives examples of various applications.

24. *Plastic Piping Systems...Here's to Your Health* shows that, based on U.S. government safety reports, the use of plastics in manufacturing plastic pipe is one of the safest methods of making pipe when compared to other non-plastic piping materials.

25. *Plastics Learn Their Role in Plumbing System Design* explores plastic pipe usage in industrial and commercial applications.

26. *The Real Green Piping Systems* explains how a myopic view of considering only one aspect of pipe selection can obfuscate the best selection of a piping material.

27. *Reduce Greenhouse Gas Emissions with Plastic Piping* proves that plastic pipe – from stages of plastic feedstock to delivered product – lessens the amount of greenhouse gas emissions to the environment when compared to many other non-plastic piping materials.

28. *Similarities of Plastic Tape and Pipe* compares the strikingly similar evolution of vinyl tape and piping products.

29. *Sustainability and Plastic Pipe* presents a step-by-step explanation of why plastics should be considered one of the most sustainable piping materials available.

30. *Sustainable Piping Systems for Green Building* is a short treatise of how plastic piping systems can assist in green building project technology by saving lives, energy, and water.

31. *Think Plastics: Thermoplastic Piping Gaining Industry Acceptance* lists reasons for the slow growth of plastic industrial piping systems in North America, and offers several reasons why more plastic piping should be specified and installed in industrial applications.

32. *Trenchless Piping and Plastics* offers three proven methods of rehabilitating and installing new underground plastic piping systems with minimal disruption of above ground activities.

33. *Why Plastic Piping* gives several reasons for the meteoric growth of plastic piping throughout the world.

ADVANTAGES OF FLEXIBLE GROOVE-COUPLED
PLASTIC PIPING SYSTEMS

Most mechanical engineers and pipe fitters understand the benefits of using plastic piping systems. Plastic piping has a long list of advantages: it is lightweight, as well as corrosion and chemical resistant; it has optimum flow characteristics and also low thermal conductivity; it is exceptionally durable, environmentally sound, as well as cost-effective; and just as important, it is easy and safe to install. This last benefit is worth a closer look. In this article, I will discuss a proven but seldom used joining method for plastic piping installations.

The most common methods of joining aboveground plastic piping systems are either heat fusion or solvent cementing, depending on the piping material. Types of plastic joined by solvent cementing are Acrylonitrile Butadiene Styrene (ABS), Chlorinated Polyvinyl Chloride (CPVC), Polyvinyl Chloride (PVC), and Polystyrene (PS). Types joined by heat fusion are Polyethylene (PE), Polypropylene (PP), Polyvinylidene Fluoride (PVDF), and other fluoropolymers. Solvent cementing and heat fusion joining methods produce excellent joint integrity and strength, producing joints capable of handling the same or even higher working pressures as the rating of the piping products being joined.

It is also important to know that there are other plastic pipe joining methods—mechanical ones—that are available when conditions of service allow; working pressures up to 150 psi or potentially higher depending on the pipe diameter and product design are available. These mechanical methods include: threading, flanging, and flexible grooved couplings. Whether your piping systems are joined by any of these joining techniques, all offer similar advantages: they are able to be assembled, disassembled, and potentially reused; less training is necessary to successfully install; installation is not normally delayed by job site weather conditions; frequently these joining methods have lower installation costs; and mechanical systems are easy to maintain. This article will concentrate on one of these alternate joining methods—flexible grooved couplings.

Grooved piping first appeared in Europe during World War I and was developed to facilitate rapid deployment of fuel and water lines for Allied forces. This *victory piping*, as it was called, has been in use for almost 100 years. Grooved mechanical coupling has four components: grooved-end pipe, a two-part coupling housing, an elastomer gasket, and bolts and nuts. The pipe ends are grooved by either roll or machined grooving at the job site, contractor facility, or the pipe manufacturer's plant. The key sections of the coupling housing engage the grooves.

Figure 1
Machine roll-grooving pipe.

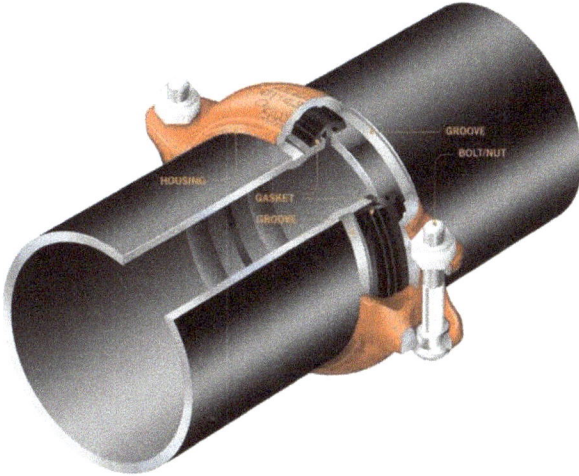

Figure 2 Components of flexible grooved coupler.

The specially designed gasket is resilient and pressure responsive. The bolts holding the housing segments together are tightened with the only joining tool required—a wrench. In the installed state, the coupling housing encases the gasket, and it engages the groove around the circumference of the pipe to create a unified joint that provides a leak-tight seal in a fully self-restrained pipe joint.

Figure 3 Joining sequence of flexible grooved piping

Once the joint is installed and tested, it requires little or no maintenance for the life of the system. In normal use, the gaskets do not require any periodic lubrication or replacement. The gasket and housing design keep the nuts and bolts from loosening over time if subjected to vibration from pumps or other equipment, or as a result of axial movement caused by thermal changes.

Currently, most grooved joining is metal based using aluminum, copper, ductile iron, galvanized steel, or stainless steel piping. There have been, however, dozens of successful grooved joining installations using mainly PVC and, to a lesser extent, PE piping product. There is little history of other grooveable plastic piping materials yet, but with additional testing and research, there is no reason that ABS, CPVC, PP, and PVDF piping could not qualify for grooved joining technology.

When choosing a grooved piping system, it's important to always follow the pipe and coupler manufacturer's recommendations and instructions. For example, with PVC pipe, only certain pipe diameters and wall thicknesses may be used in conjunction with a specific type of groove—cut or rolled. All plastic piping installations use a flexible (as opposed to a rigid) coupler assembly.

What about grooved plastic fittings and valves? Plastic piping can be joined with standard grooved metal products as in Figure 4. But simple fabrication techniques, standard injection molded plastic fittings and valves can be easily adapted to a grooved piping system. ABS, PVC, and CPVC piping socket-end products can be transitioned into a grooved system using spigot-grooved adaptors. The spigot-adaptor, with an appropriate cut or rolled grooved end, is constructed from pipe to the exact socket depth of the fitting and/or valve to be joined. To convert the socket-molded products to a grooved system, simply solvent cement the fabricated spigot adaptors into the fitting/valve socket and, voila, you now have an all plastic grooved pipe, fitting, and valve system.

Figure 4 PVC flexible grooved-pipe installation.

With heat-fused plastic piping such as PE, PP, or PVDF, similar adaptors may be used with socket fittings/valves, but instead of solvent cementing the adaptors in the product socket, heat fusion joining methods are used. Many heat-fused systems use butt fusion technology for joining in which the fittings/valves have the same diameter and wall thickness of the pipe. In this case the fitting/valve can be directly grooved and joined to the pipe negating the need for adaptors of any kind.

There are dozens of advantages for using flexible grooved-coupled plastic piping systems.

Table 1 lists some of the major features and benefits of these systems.

Table 1
Features and Benefits of Flexible Groove-Coupled
Plastic Piping Systems

Feature	Benefit
No flame, hot-plates, or solvents	Reduces injury and less environmentally intrusive
No expensive joining tools	Reduces installation costs
Just two bolts to tighten or loosen	Ease of assembly and disassembly
No cure time required after assembly	System can be tested immediately after joining
Less training required of installers	Easy, safe, and cost-effective to join
No drainage required for repairs	Saves downtime and process fluid loss
Piping can be rotated 360°	Speeds up installation of fittings and valves
No weld x-ray to test joints	Leaks in testing can be seen and corrected quickly
Can join dissimilar pipe materials	Easy and cost-effective to join other piping systems
Piping products can be reused	Reduces future cost and is environmentally sound
Isolates noise and vibration	Minimizes system noise and vibration transmission
Provides expansion and contraction	May reduce costly expansion joints, loops, and offsets
Flexible couplers allow deflection	Minimizes system stresses
Corrosion and chemically resistant	Minimum or no protective coatings are required
Not affected by weather conditions	Reduces installation time

Just recently, an American flexible grooved coupling manufacturer introduced a coupler made from an engineered composite material. Presently, the product can handle maximum pressures up to 150 psi and is available in pipe diameter sizes of 1½ to 4 inches. Larger diameter couplers are planned for future use. The distinct advantage of using this new plastic coupler is to deliver a piping system that can withstand both interior and exterior chemical and corrosive attack in most fluid environments.

Figure 5
Engineer composite
flexible coupling.

When projects are planned in such markets as Heating-Ventilating-Air Conditioning, Plumbing, Mining, Oil & Gas, Maritime, Power, Water and Waste Water, Chemical Process, and others, it is time to consider the use of a flexible grooved plastic piping system—a proven system that provides significant features and benefits to end-users and installing contractors alike.

Note: The author would like to acknowledge the Victaulic Company, Inc. for their assistance in providing photos, reviewing, and improving the article's content. Any errors that may occur in the article are the author's.

15

THE EVOLUTION OF PIPE

It's only when the hunter-gatherers decided to change their nomadic ways and stay put did the idea and solution sprout up on how to quench the population's thirst and water their crops. At first, directional, open-slopped ditches were dug using gravity flow from a nearby water reservoir such as a well or river. Later, when the outside elements made in-soil conveyance difficult to maintain, aqueducts were built using stone and mortar. Examples of this method of water transportation are still visible in Europe and Turkey thanks to the Romans who built these durable engineered structures a few thousand years ago. But due to water evaporation, the ruining of arable land with these massive above ground structures, and the labor to build the aqueducts, pipe was invented.

Figure 1 Roman-built aqueduct.

By definition, pipe is a "tube or hollow cylinder for conveyance of a fluid." It seems the Romans again led the charge by discovering folded-seam lead piping. This product was fairly easy to produce due to its malleability and availability. In fact, lead piping was so popular for hundreds of centuries that the Latin word for lead—"plumb"—became the word root for today's installers of water piping—"plumbers." Lead water pipes were used until the early 20th century when the health hazards of lead became more fully understood.

Wood pipe hollowed out of logs came into use in the 1800s in areas of the world where lumber was inexpensive (wood piping was used to a large extent in the Northeast United States and Canada). Next, due to being more cost effective and having better performance, there was an influx of metal piping systems. Cast iron piping started making inroads in the early 1900s and was followed by other ferrous materials as well as bronze, brass, copper, and aluminum. Depending on the application, these metal piping materials are still in use throughout the world. Other piping materials developed in the 1900s, such as clay and cement, are still used in many places but mostly for underground, non-pressure applications.

Figure 2 Wooden pipe.

For the same reasons that metal pipe replaced its prede-cessors—cost and performance—thermoplastic pipe became the material of choice beginning in earnest in the 1960s. Polyvinyl Chloride (PVC) was one of the earliest plastic piping systems to handle fluids; today, it is one of the most commonly used piping materials in the world. Polyethylene (PE) is the next most voluminous plastic pip-ing material with other commonly used plastic piping mate-rials such as Acrylonitrile-Butadiene-Styrene (ABS), Polypropylene (PP), Chlorinated Polyvinyl Chloride (CPVC), and Cross-linked Polyethylene (PEX). For piping applications requiring higher temperature and broader chemical resistance capabilities, fluoropolymers such as Polyvinylidene Fluoride (PVDF) and Polytetrafluoroethy-lene (PTFE) are used.

So why have the world's architects, design engineers, installers, and end-users embraced plastics? Cost-effective-ness and performance. Today's market place has recognized

Figure 3
Variety of plastic pipe.

that plastic piping systems are easy and safe to install, reliable and long-lasting, and environmentally sound. As Darwin so rightly predicted, species sustain themselves and rise to the top of their chain by having the ability to adapt to their changing environment. This principle is also true regarding piping materials. Although there is no "perfect" piping material, probability favors plastic piping and its constantly evolving nature to maintain its place as one of the preferred piping materials in many applications today and in the future.

Reprinted with permission of the IAPD; issue october/november 2007 – **the IAPD magazine**

FAIRY TALE OR REALITY

Once upon a time there was a blind study; its purpose was to determine which of two products was greener in the field of agricultural food production: Product H or Product T? The judges were an elite panel of experts, PhDs, whose specialty was developing and publishing environmental impact studies.

Figure 1 Horse driven plow.

Advantages of Product H:
• Dozens of centuries in use
• Lower purchase cost
• Lower energy costs
• Sustainable source of supply
• Energy to operate with non-fossil fuels
• Minimum greenhouse gas emissions
• Finished product is recyclable

Figure 2 Tractor.

Advantage of Product T:
• Cost effective

The unanimous decision of the panel was predictably Product H, the horse-driven plow, When farmers were polled, their choice was Product T, the tractor.

This tongue-in-cheek story might seem farfetched, but it illustrates the flaw of "not seeing the forest for the trees"—an apt, if timeworn phrase. The same flaw in thinking is apparent today from "experts" as well as vocal but misinformed environmental acolytes, who insist that plastics are detrimental to the planet, have no sustainable benefits, and can be easily replaced with other materials. They are missing the broader view: plastics offer not only environmentally sound, but also cost-effective solutions in an ever-growing and extensive list of applications.

Let's examine the field of plastic piping. Just like the farmers from the story who chose the tractor on the basis of cost-effectiveness, the domestic and international marketplace has chosen plastic pipe in most cases over other materials for applications that conditions of service allow.

Why are plastics the preferred piping material in many applications such as residential and commercial hot and cold-water distribution, natural gas distribution, acid-waste drainage, residential drain-waste-vent, irrigation, swimming pools, water mains, sewer lines, central vacuum systems, radon elimination, rainwater harvesting, residential fire suppression systems, geothermal energy, water well casings, ultra-high purity water systems, aquariums, water theme parks, fountains, and several others?

Why has the marketplace embraced plastic piping? Four reasons: for its durability, safety and ease of installation, environmental soundness, and cost-effectiveness. Top ratings in these four categories make plastic one of the best bargains in the fluid-handling industry.

Figure 3 Pallets of PVC pipe.

1) Durability

Compared to other piping materials, plastic piping is almost always more durable. This is due to its resistance to chemical and corrosive attack, abrasion resistance, joint integrity, optimum flow characteristics, and constant testing to rigorous codes and standards.

2) Safety and ease of installation

Plastic piping is lightweight; it's easy and quick to make leak-proof joints. It offers a variety of joining methods, a variety and breadth of product line; ease of fabrication; multiple exterior-walled color identification, and code acceptance. All products have easy-to-decipher identification labeling; piping products are in stock and available for delivery in hours; and the product's light weight leads to reduction of on-site injuries while speeding up installation time.

3) Environmental soundness

Energy savings is one of the payoffs of plastic piping's low thermal conductivity, favorable flow rates, and reduced energy of hauling product. In addition, plastic piping is odorless and non-toxic, fully recyclable, and—based on scientifically conducted Life Cycle Assessments (LCAs)—has "greenness" that the other piping materials have difficulty in matching.

4) Cost-effectiveness

Whether it is costs for material, installation, or maintenance, plastic piping has proven time and again that few other piping materials are as cost-effective. In addition, plastic piping offers more savings by way of fewer on-site thefts, lower transportation costs and lower insurance premiums.

Summary

The moral of this story is when evaluating piping materials (or, for that matter, any other durable product), don't just myopically look at one facet of a product line. Consider the complete list of features and full benefits package so a carefully considered, well-informed, and client-focused decision can be made. You will then realize that there is no fairy tale when it comes to recognizing plastics as the overwhelming choice for most piping projects.

Reprinted with permission of the IAPD; issue december/january 2011 – **the IAPD Magazine**

HOW TO SELECT THE PROPER THERMOPLASTIC
INDUSTRIAL PIPING MATERIAL

Plastic piping is used to move more fluids than any other piping material in the world. Why? Plastics are more durable, easier and safer to install, environmentally sound, and cost-effective relative to most other piping materials. Yet, due to the many available plastic piping systems, selecting the right piping product can be a daunting task. The two keys in choosing the proper plastic piping material are knowing: 1) the complete project's service conditions and 2) the most cost-effective and suitable materials for these service conditions.

Service Conditions

The more one knows about the factors of a particular field application, the more likely one is to select the proper piping material. An information checklist for all piping installations may include the following, with the most important factors in bold print:

- *Type or purpose of application.* Application type (chemical processing, water lines, sewage, irrigation, etc.); above or below ground; any prolonged exposure to sunlight; opened- or closed-loop system; quick-closing valves in system.

- *Type of fluids to be transported.* Chemical compatibility of pipe, valves, and fittings; any special fluid data such as high particulate matter, viscosity, specific gravity, velocity, etc.

- *Maximum allowable working pressure.* Working pressure varies for each plastic piping component and may be limited to the maximum allowable working pressure for particular products such as valves, flanges, or unions.

- *Temperature range.* Temperature limitations vary for each plastic piping material and directly affect working pressure capability of all plastic products coming in contact with the transferred fluid.

- *Code restrictions.* Depending on the application, local, state, and federal codes may restrict the selection of a particular piping material.

- *Estimated life of piping system.* In very severe service conditions, various piping materials may have limited life and affect which piping system will last the longest.

- *Preferred method of joining.* Depending on the availability of utilities (electric, gas, or neither) as well as the experience of the installation crew, selection of the joining system could be important.

With the conditions of service known and knowledge of the piping material properties, specifying engineers or end-users are well on their way to selecting the best piping material for the project.

Plastic Piping Materials 1

There are five commonly used thermoplastic industrial piping materials in the United States, all of which have different temperature and allowable working pressure ranges. In most cases, all the plastic piping materials can withstand temperature lows of 0°F/–18°C. The only caveat to remember with thermoplastic piping at lower temperatures is that the piping system becomes more brittle. In the listing below, we have stated the maximum temperature capability of each piping material. Keep in mind though with thermoplastic piping, as the temperature increases, the allowable working pressure decreases. Therefore, in most cases, only drainage applications can be used at the maximum temperature rating.

- *CPVC (Chlorinated Polyvinyl Chloride).* Similar in properties to PVC; can handle temperatures up to 210°F/99°C; used in pressure and chemical drainage piping systems as well as hot and cold water distribution and fire sprinkler systems   .

- *PE (Polyethylene).* Several material compound types depending on the application; the broadest range of pipe diameters; mostly used in below ground applications; most PE piping systems have temperature capabilities of up to 160°/71°C; however, one fairly recent material, polyethylene raised temperature (PE-RT), can withstand temperatures up to 210°F/99°C; a preferred piping material for natural gas distribution.

---

[1]Acrylonitrile Butadiene Styrene (ABS) and Crossed-linked Polyethylene (PEX) are not shown because most of these excellent plastic piping materials are used in the United States for residential/commercial, not industrial applications.

- *PP (Polypropylene).* Very versatile material for temperatures up to 180°F/82°C; the leading piping material in chemical drainage and, because of its excellent chemical resistance, can handle many corrosive pressure piping applications.

- *PVC (Polyvinyl Chloride).* When conditions of service allow, PVC is one of the most specified and installed plastic piping material in the world; maximum temperature up to 140°F/60°C; in most cases PVC is the lowest cost piping material for a myriad of applications.

- *PVDF (Polyvinylidene Fluoride).* Has the best chemical resistance properties and highest temperature capability—275°F/135°C—of any of the listed piping materials; mostly immune to ultra-violet degradation; the preferred piping material in ultra-high purity water systems.

Examples

Each of these plastics has properties that may or may not be suitable for a particular application. To determine the proper piping material, we need to compare the project's conditions of service with that of the piping material's properties, breadth of product line, and installed costs. The best way to show how this process works is by example.

Example: Service Conditions

The example's listed service conditions that follow are the basis for selecting piping materials:

- Application. Metal ore smelting; above ground
- Fluid. Sulfuric acid with 70% concentration
- Working pressure. 50 to 105 psi
- Working temperature. 70°F/21°C to 130°F/54°C
- Piping size. ½ to 6 inches

Example: Chemical Compatibility

The first step is to determine the chemical resistance of the fluid versus the piping material. There are several published and well-tested chemical resistance tables that list hundreds of chemicals and their suitability to both plastics and metal materials. Table 1 shows a snippet of a typical chemical resistance chart pertinent to our example. Upon reviewing the chart, you can see that PP and PE piping are not able to handle the conditions of service shown.

Table 1

Partial Chemical Resistance Chart

Chemical	PVC		CPVC			PP			PVDF			PE	
Temperature (°F)	70	140	70	140	185	70	150	180	70	150	250	70	140
Sulfuric acid, 50%	R	R	R	R	R	R	R	NR	R	R	R	R	—
Sulfuric acid, 60%	R	R	R	R	R	R	R	NR	R	R	R	R	—
Sulfuric acid, 70%	R	R	R	R	R	R	NR	NR	R	R	R	R	—
Sulfuric acid, 80%	R	R	R	R	R	R	NR	NR	R	R	—	R	NR
Sulfuric acid, 90%	R	NR	R	R	R	R	NR	NR	R	R	—	R	NR
Sulfuric acid, 93%	R	NR	R	R	R	R	NR	NR	R	R	—	R	NR
Sulfuric acid, 100%	NR	NR	NR	NR	NR	NR	NR	NR	NR	NR	—	NR	NR

R = Recommended   — = No information available   NR = Not Recommended

Example: Pressure Capability

Next we determine the working pressure capability of the piping materials in the example. Thermoplastic piping working pressure is affected by temperature and, in the case of schedule 40 and 80, piping can vary with pipe diameter. Tables 2 and 3 show the working pressure capabilities at ambient temperature (73°F/23°C) of the five piping materials and the temperature correction factors that are used to multiply the factor times the ambient pipe pressure rating in order to obtain the maximum working pressure capability of each piping material. As noted in the tables, PVC 6-inch pipe at ambient temperature has a working pressure of 280 psi. The temperature correction factor for PVC pipe at 130°F/54°C is shown as 0.30. Therefore, multiplying 280 psi by 0.30 equals 84 psi. The conditions of service list a maximum working pressure of 105 psi. Hence, only CPVC and PVDF piping are capable of handling the stated conditions of service.

Table 2

Maximum Operating Pressure Ratings (psi) of Schedule 80 Piping / Fittings @ 73°F/23°C				
Nominal Pipe Size (in.)	PVC / CPVC	PE (SDR 11)	PP*	PVDF
3	370	160	190	250
4	320	160	160	220
6	280	160	140	190

Table 3

Temperature Correction Factors for Piping					
Operating Temp. (°F/°C)	CPVC	PE	PP	PVC	PVDF
110/43	.77	.74	.80	.50	.75
120/49	.70	.63	.75	.40	.68
130/54°C	.62	.57	.68	.30	.62

✓ CPVC: 0.62(280) = 174 psi

✗ PVC: 0.30(280) = 84 psi *(System max. is 105 psi)*

✓ PVDF: 0.62(190) = 118 psi

Example: Cost Comparison

The last step in the selection process is to choose the more cost effective of the two qualifying piping materials. Looking at Table 4 it is evident that CPVC is the winner—the PVDF piping system costs over two and a half times that of CPVC.

Table 4

Estimated Installed Costs of Six-Inch Diameter Piping Systems*	
Materials	Piping Costs ($)
PVC Schedule 80	11,499
PE SDR 11	14,374
CPVC Schedule 80	18,932 ✓
Carbon Steel Schedule 40	22,789
Stainless Steel 304L Schedule 40	45,835
Stainless Steel 316L Schedule 40	49,355
PVDF Schedule 80	63,945 ✗
*PE and PP cost estimates by the author	

There are industrial applications where temperatures exceed 275°F/135°C and working pressures are well above 250 psi. For these conditions of service, piping of glass-reinforced thermosetting resin and metal- and plastic-lined metal may be the materials of choice. But for an estimated 80 percent of all industrial applications, thermoplastic piping systems is normally the preferred choice due to their durability, ease of use, environmental soundness, and cost effectiveness.

Note: The example and charts shown are adapted from the *Thermoplastic Industrial Piping Systems Workbook* published by the Plastic Pipe and Fittings Association (PPFA)

Reprinted with permission of the IAPD; issue february/march 2013 – **the IAPD magazine**

18

THE IDEAL PIPING SYSTEM

Suppose a group of engineers and installers got together to collaborate on designing the ideal piping system; they were aiming for an overall piping solution to meet the requirements of residential, commercial, and industrial building projects. The requirements: it would have to be durable, easy and safe to install, environmentally sound, and, not least, cost-effective. Plastic piping makes the grade on all counts.

Figure 1 Plastic pipe lightweight reduces on-site accidents and reduces installation time.

Durability

Plastic piping has a proven performance record for over seven decades. Many installations have a service life projected to exceed a hundred years. Plastic piping is practically immune to interior and exterior corrosion or chemical attack. It doesn't rust, pit, or scale. It resists bacteria, fungi, and termite attack. It has unmatched leak-proof joint integrity. It's more abrasion resistant than most other piping materials. It has a smooth inner pipe bore offering optimum flow characteristics. It's non-toxic and odorless. And it's available in complete systems of fluid and air-handling products.

Figure 2
Plastic piping has excellent chemical resistance, as shown in this saltwater aquarium piping.

Easy and Safe to Install

When it comes to the ease and safety of installation, plastic piping is in a league of its own. Plastic's weight (1/6 to 1/8 the weight of most other non-plastic piping) makes installation comparatively easier and safer. Along with its lighter weight, its ease of joining also reduces the likelihood of reportable on-site accidents. Adding value is that there are over a half-dozen proven joining methods, depending on the application and particular plastic piping material. Plastics are available in a broad size range, from 1/8-inch to 144-inch pipe diameters and larger. Generally, from 10-inch diameter and smaller, no heavy pipe moving equipment is required. Process-integrated coloration allows piping to be easily identified, especially important in critical applications, and service markings allow for easy identification of all relevant data. Plastic pipe's low thermal conductivity minimizes or eliminates insulation. Another benefit of plastic piping is its acceptance by most building code authorities throughout the world.

Environmentally Sound

All plastic piping compounds used for potable and high purity water applications are produced to National Sanitary Foundation International and ASTM standards. Presently, Life Cycle Assessments have been completed in Europe and underway in North America to scientifically determine the environmental impact of plastic pipe and other piping materials. The preliminary results so far are very complimentary for plastic. Part of the analysis shows plastic pipe reduces energy losses due to low thermal conductance. With its smooth inner pipe bore, plastic piping requires less horsepower to move fluids. Helping to further reduce its environmental footprint are additional energy savings in shipping costs. In post-industrial processing, almost 100% of the material is recycled. In post-consumer use, all thermoplastic piping is recyclable. Unlike any other non-plastic piping materials, plastic piping in the future could be made from sustainable and renewable bio-fuel feedstocks.

Figure 3 Plastic pipe's smooth inner surface reduces friction loss.

Cost Effective

So here we have a piping system which possesses amazing durability, is easy and safe to install, and is environmental sound. What more could we want? How about cost-effectiveness? Plastic piping systems reliably cost less to purchase, install, and maintain than almost any other piping systems. And when you consider the cost savings in freight and reduced onsite theft, plastic piping is at the head of the pack.

When designing and selecting a piping system, make the smart move. Choose plastics—*the ideal piping system.*

Reprinted with permission of the IAPD; issue june/july 2010 – **the IAPD magazine**

JOINING OF THERMOPLASTIC INDUSTRIAL PIPING SYSTEMS

Ask people who have experience with thermoplastic industrial piping systems—the specifying engineers, installers, and end users—why plastic is their material of choice whenever conditions of service allow. They'll be quick to tell you about plastic's durability, the ease and safety of installation, environmental soundness, and cost-effectiveness.

Another amazing benefit of plastic piping just as important to know is the versatility of joining options available for installation. There are at least three different joining methods for each plastic system. This article briefly describes several of these methods.

Solvent welded, heat fusion, and mechanical coupling are the major categories of joining plastic pipe. Solvent welded and heat fusion are unique to plastic piping. The joints resulting from these two joining methods normally have the same or higher working pressure ratings than either of the products to be joined has alone. The only way to disconnect products joined by cementing and fusion is to destructively cut out the troublesome joint.

Solvent Welding

Solvent welding uses a cementing fluid that contains solvents to dissolve or soften the surfaces being bonded so that the bonded assembly becomes essentially one piece of the same type of plastic. A proper solvent-welded joint is sometimes referred to as a *homogeneous monolithic joint*. In other words, if you took a cross-section of the cemented joint, the two bonded areas would appear as one solid layer of material.

The solvent cementing process requires simple and inexpensive tools to cut, deburr, bevel, and clean piping products to be joined. Then using a suitable applicator (if using a brush, use a brush width half the diameter of pipe), coats of primer (when required) and cement are applied to the pipe exterior and female fitting socket. While still wet, the pipe is inserted into the female socket with a 1/4 turn and held for several seconds to prevent pipe push-out. Any excess cement is then removed by a rag. Many in the plumbing and pipe fitting industry refer to solvent cementing as *gluing*. Nothing raises the hackles of plastic piping purists as when someone states plastic pipe is "glued" together.

Plastic piping materials that can be joined using solvent cementing are: Acrylonitrile Butadiene Styrene (ABS), Chlorinated Polyvinyl Chloride (CPVC), Polyvinyl Chloride (PVC), and Polystyrene (PS).

Figure 1
After cutting and beveling PVC pipe, wipe off dirt and any other foreign substance.

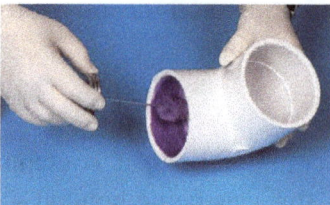

Figure 2
Using appropriate dauber or brush, apply primer to fitting socket.

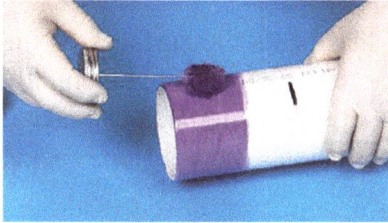

Figure 3
Apply primer to outside pipe wall.

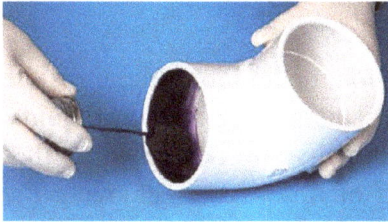

Figure 4
Before primer dries, apply cement to fitting socket.

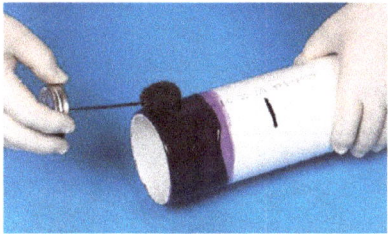

Figure 5
Before primer dries, apply cement to exterior pipe wall.

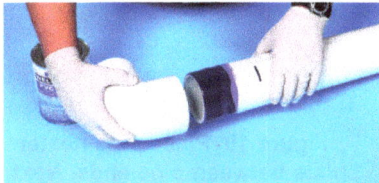

Figure 6
Quickly join pipe and fitting, giving pipe a quarter turn when bottoming out pipe to fitting stop.

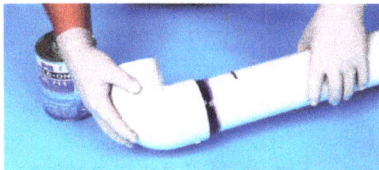

Figure 7
Hold fitting and pipe together for 10 to 15 seconds to prevent any back out.

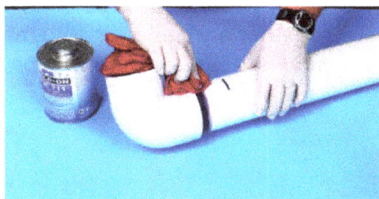

Figure 8
Wipe off any excess cement after joint is completed.

Heat Fusion

Heat fusion is a process that heats the surfaces of the parts to be joined so that they fuse and become essentially one piece, with or without additional material. Let's take a look at several types of heat fusion: butt fusion, socket fusion, and electro fusion.

Butt Fusion

The most popular heat fusion method for joining fusible industrial pipe is butt fusion. This joining method requires clean and deburred piping ends that are heated with a hot plate for a prescribed period of time. After removing the plate, the ends are then pressed together by a manual or hydraulic pressure device.

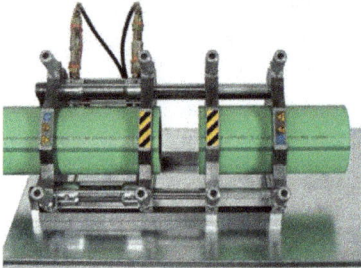

Figure 9 Cut pipe and clean ends; then place and align pipe ends into the

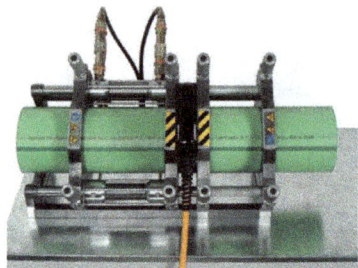

Figure 10 Place pipe heater plate between pipe ends. Next, push pipe ends to the plate for a listed period of time.

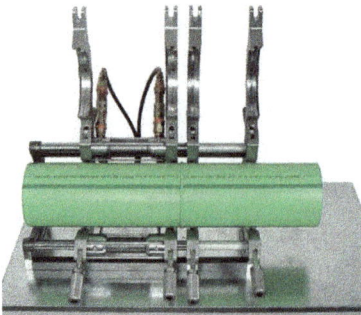

Figure 11 After removing heat plate, push pipe ends together and hold for several seconds until a uniform pipe bead is visible.

There are specialized butt fusion joining methods that use infrared radiation and bead- and crevice-free technologies especially designed to handle ultra-high purity water systems for the semi-conductor industry. The infrared fusion method prevents possible contamination from metal hot plate micro-particles; the bead- and crevice-free method minimizes flow turbulence and sources for bacteria buildup.

Figure 12 PVDF pipe being joined in a clean room using infrared machines.

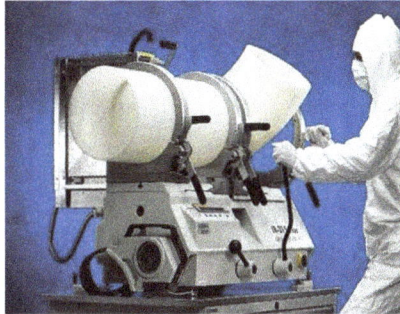

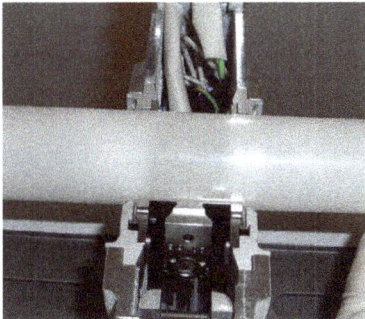

Figure 13 PVDF finished bead- and crevice-three joined.

Socket Fusion

Smaller diameter piping can be butt fused, as well as socket fused. Socket fusion uses the same principle as butt fusion, but instead of joining the pipe ends, the outside diameter of the pipe is heated simultaneously with the interior fitting socket. After a prescribed period of time, the heat tool is removed and the softened pipe end is inserted into the fitting socket using manual or machine force of pressure; it is then allowed to bond over several seconds.

Figure 14 After cutting, deburring pipe end, and cleaning areas to be joined, place pipe and fitting into the proper heated pipe/fitting dies, and hold until the pipe and fitting socket soften.

Figure 15 Pull softened pipe and fitting from heated dies.

Figure 16 Push pipe into fitting socket and hold for several seconds (do not twist pipe).

Electro Fusion

The third type of heat fusion used to join plastic piping is electro fusion. This method uses copper coils that are either manually inserted into the fitting socket or that are already imbedded by the manufacturer into the fitting socket. The pipe is then inserted into the fitting, and the electric coil wires are connected to a device that generates an electrical charge to heat the coil to the proper temperature. The heat from the coil joins the two surfaces together. Typically low working pressure or drainage systems use this fusion technology.

Figure 17 Turn on and heat the electro fusion machine.

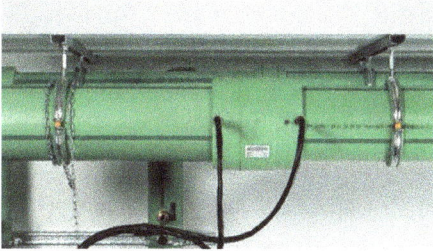

Figure 18
Machine leads attached
to fusion-fitting
electric coils.

Piping materials that use all these heat fusion joining methods are: Polyethylene (PE), Polypropylene (PP), and Polyvinylidene Fluoride (PVDF). (Note: At the present time, there is one U.S. pipe manufacturer of a specially formulated PVC compound that offers a butt-fused underground piping system; all other PVC piping systems for underground use are either solvent-cemented or use a gasketed-bell and spigot joining technique.)

Mechanical

Almost all industrial plastic piping systems can be flanged, threaded, and mechanically coupled. These systems can couple piping of similar or dissimilar piping materials and, for the most part, are easily disconnected for repair, replacement, or disassembly. Plastic flanges are either fixed-ring or Van Stone type (ring rotates around the adapter). Depending upon the pipe material, the flange can be directly joined to the pipe or a backup ring of plastic or metal can be inserted first, followed by a pipe-end adapter either cemented or fused to the piping products.

Figure 19
PVC flanging using a
torque wrench to
tighten bolts.

Direct threading of plastic piping using specialized tools and threading dies is possible. Depending upon the plastic pipe material, however, direct threading reduces the working pressure of the pipe by half with PVC and CPVC Schedule 80 and 120 pipes (thinner-walled PVC pipe is not recommended for threading). Direct threading of other plastic piping materials reduces the pipe's working pressure to twenty-five psi or less. A better method of implementing a plastic threaded system is to use threaded adapters that attach to the plastic piping using solvent welded or heat fusion joining methods. Keep in mind when using flanges, unions, or threaded adapters, the maximum working pressure of the piping system in many cases is limited to 150 psi.

Figure 20 Plastic by threaded metal transition union.

Another mechanical joining system mostly used for larger diameter PVC thicker walled piping is the cut- or roll-grooved type. This joining method uses a tool to cut or roll-groove pipe ends that are joined using a metal compression coupling with an internal rubber seal that clamps and holds the piping together in a leak-free manner.

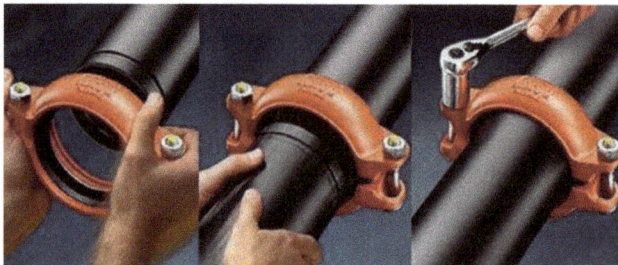

Figure 21 Metal sealing coupler joining grooved pipe.

As shown, there are a variety of well-proven ways to join industrial plastic pipe. Choosing the most durable and cost-effective joining system should be based upon the piping material, the experience of the installing crew, and the field service conditions of the site. For more detailed information on the joining of thermoplastic piping materials, go to the following website: www.pphahome.org

With the many features and benefits of plastic piping, including the variety of joining methods, it's no wonder plastics are one of the fastest growing piping materials in the world.

20

JOIN THE REVOLUTION...THINK PLASTICS

The famous 19th-century American physician and author, Oliver Wendell Holmes, stated it best when he wrote, "A mind that is stretched to a new idea never returns to its original dimension." And what is this revolutionary new idea? The use of thermoplastics in industrial piping applications.

Wood, clay, concrete, and metal piping are just some of the piping materials in use for centuries. But with today's modern technology there is a better piping alternative material—thermoplastics. This article will list the many advantages and benefits of plastic piping, but the bottom line is Thermoplastic Industrial Piping Systems (TIPS) are more cost-effective in almost every industrial piping application compared to alternative piping systems. Yet in the industrial piping market, where TIPS is capable of handling an estimated 70% of all applications less, than 15% is actually used.

In the last four decades, plastic have become one of the dominant materials in many piping markets including residential drain-waste-vent, natural gas distribution, acid waste drainage, water lines, underground irrigation, swimming pools, and water-theme parks. However, the bulk of industrial and commercial markets have been slow to embrace the use of thermoplastics. Why?

Figure 1
PVC roof drainage piping
for commercial building.

First, the plastics industry must share some of the blame for not having done a better job of educating the market place to the benefits and capabilities of TIPS products. This oversight is being addressed presently, in part, by the Plastic Pipe and Fittings Association (PPFA) product line committee-TIPS as well as special plastic piping distributors in concert with the International Association of Plastic Distribution (IAPD).

Second, minimal research and development of TIPS products have been performed in the United States. As a result, few innovative products or piping materials have been developed domestically; most new developments are coming from off-shore manufacturers.

Third, U.S. industries and institutions—in conjunction with labor unions and building code associations— have been to a large degree unwilling or very slow to change their habits and adapt to more progressive and efficient piping materials. Older industries such as pulp and paper, electric utilities, oil and gas, petroleum, metal refineries, and commercial/institutional buildings could greatly benefit from increased use of TIPS. Also a lack of economic pressure in past years has kept these industries from moving toward plastic piping materials. But today's highly competitive global environment will most likely make change easier. By

American industry joining the revolution in the use of thermoplastics (one of the most effective and cost saving piping materials), domestic products will become more competitive in foreign markets, increasing U.S. exports and adding more jobs.

Why Plastics?

The advantages and benefits of plastic piping are significant and will finally turn the manufacturing industry in favor of its use. Thermoplastic features can produce considerable cost savings while increasing piping system reliability. Advantages include:

Corrosion resistance. Plastics are nonconductive and, therefore, immune to galvanic and electrolytic corrosion. Plastic piping materials are so corrosion-resistant that they can be buried in alkaline or acid soils or installed in aboveground environments with minimum protection required.

**Figure 2
Corroded metal piping.**

Lightweight. Most plastics are a minimum of 1/6 the weight of metal piping. This means less freight costs. The benefit of a lightweight material allows easier installations in close quarters and doesn't require expensive lifting equipment.

Optimum flow rates. The interior wall of almost all plastic pipes has a Hazan and Williams C Factor of 150 or higher. This means less energy or horsepower is required to transfer fluids or smaller diameter piping may be used resulting in cost savings.

Figure 3 Plastic pipe smooth inner wall.

Low thermal conductivity. Most plastic piping has lower thermal conductance, which means more uniform temperatures when transporting fluids. Minimal heat loss through the pipe wall of plastic piping may eliminate or greatly reduce the need for piping insulation.

Chemical resistance. The variety of common plastic piping materials allows most chemicals at moderate temperatures to be successfully handled. The plastics industry has a listing of hundreds of chemicals that may be compatible with a given plastic. This helps to eliminate the guess work for end users and engineers.

Variety of leak tight joining methods. Plastic piping can be joined in several leak proof ways including being solvent welded, heat-fused, threaded, flanged, and mechanically coupled. These methods allow easy joining and adaptability to other non-plastic piping materials.

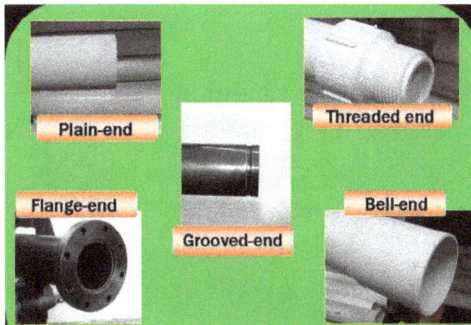

Figure 4 A few joining methods of plastic piping.

Abrasion resistance. The molecular toughness and inner bore smoothness of plastic pipe makes it ideal for abrasion-resistant applications such as fly ash and bottom ash as well as many other abrasive slurries and solutions.

Color variety. The plastic piping extrusion process allows color to be an integral and homogenous part of the piping. No external painting is required. Vibrant colors are especially important and available for underground installations, so as to be highly visible when contractors are excavating. That way, the contractors can easily see the pipe, preventing pipe damage and minimizing any safety concerns.

Piping system integrity. Most plastic piping systems come complete with complementary products such as pipe, valves, fittings, tanks, and pumps. This feature allows complete systems of one plastic material to be in contact with all fluid wetted parts (especially for cemented and heat-fused joined systems).

Maintenance free. A properly installed plastic piping system requires minimum or no maintenance. It's as simple as that. There is no rust, pitting, or scaling; no galvanic or electrolytic corrosion; external pipe coatings are minimized or not needed; and buried plastic piping is not affected by even the most aggressive soil conditions.

Flexibility. Thermoplastic piping materials are relatively flexible compared to other piping materials. This feature, coupled with optimum flow rates, allows some plastic pipe to be used as insertion liners in failing non-plastic piping. Also, the flexibility of plastic piping in underground piping reduces the use of fittings with the allowable bending radius in plastic pipe to as little as 20 times the outside diameter of the pipe.

Codes. There are dozens of plastic piping standards used or referenced in Building, Plumbing, Mechanical, and Electrical Codes, and AWWA, FM, NFPA, AGA, EPA, DOT, DOD, and API publications. These standards ensure that thermoplastic products maintain uniform characteristics, which in many cases, allows each manufacturer's products to be used interchangeably with others.

Costs. When all of the above features are considered, significant savings almost always result when using plastic piping systems compared to other piping systems and includes savings in: product purchases, labor, maintenance, lack of job site thefts, and transportation.

Applications

There are very fluid handling applications in which plastic piping is limited in getting the job done. The following examples serve to demonstrate the varied use of TIPS in many industries and markets:

Food Processing. Most plastic piping materials are approved by the National Sanitary Foundation and receive Food and Drug Administration approval when required. The purity of the end product in any food-processing application is critical and plastics fit the bill beautifully.

Surface Finishing. The automotive, aircraft, electrotyping, and canning industries use TIPS where possible in their plating processes. Plastics are a natural in this market, since almost every metal-salt plating solution can be handled easily including brass, cadmium, chrome, copper, gold, lead nickel, rhodium, silver, tin, and zinc plating processes.

Figure 5 Bleach process using plastic piping.

Steel Mills. Ironically, steel mills are replacing steel piping with plastics. The mills realized that their manufacturing costs improved with use of plastics because of reduced maintenance, lower material costs, and longer life provided by TIPS.

Pulp and Paper. These plants handle four types of media: liquids, steam, water, and stock. Except for steam, plastic piping can handle most of the other fluids under 275 °F/135°C and 150 psi.

Electronics. The manufacturers of solid-state electronics products such as semiconductors, rectifiers, and printed circuitry demand ultra-pure water to clean their products and prevent contamination. Thermoplastics are the preferred materials for handling ultra-pure water in which an ion exchange or demineralization system is employed. PVC, CPVC, PP, PVDF, and other fluorocarbons are used for water distribution systems. TIPS also are used for handling etching media such as sulfuric, nitric, hydrochloric, and hydrofluoric acids. Wastewater and air-handling systems required in the electronic industry also use TIPS throughout.

Figure 6
PVDF piping system in
pharmaceutical plant.

Mining/Oil/Gas. Plastic pipe, itself a derivative of oil and natural gas, has successfully been applied in handling most crudes, salt water, and natural gases. Most natural gas distribution today uses millions of feet of plastic pipe. Polyethylene piping—colored beige, yellow, or orange—is the preferred material for this application. In the mining industry, the most popular use of thermoplastics is in ore leaching, in which the ore is treated with dilute sulfuric acid or sulfides, and then with ferric sulfate solutions. PVC, CPVC, ABS, and PE piping are used in many of the leaching process stages. Plastics also are used for the movement of ore slurries and other piping applications in under and above ground mining.

Marine Applications. Shipbuilding, marinas, fish hatcheries, marine research, aquariums, and theme-water parks are using significant amounts of plastic piping especially in salt water environments where internal and external corrosion resistance is important.

Water/Waste Treatment Plants. Whether in primary or secondary treatment phases, plastics are used throughout water and sewage treatment facilities. Influent and effluent lines, sludge lines, chlorine and sodium hypochlorite lines, fluoride and alum lines, and many other piping lines use TIPS.

Heating/Air conditioning/ Refrigeration. Millions of feet of PVC and CPVC piping have been used in central air conditioning systems of institutions and commercial buildings, for condensate return lines, to handle brine solutions in refrigeration processes, and for the refrigeration process in ice skating rinks (polyethylene piping).

Figure 7 Plastic pipe in water/waste treatment plant.

Institutional Facilities. Hospitals and school complexes are large users of TIPS particularly in acid-waste drainage lines for chemistry, physics, and hospital laboratories. The plastic piping materials used for this application are PP, CPVC, and PVDF.

Figure 8 PE coiled natural gas distribution piping.

Additional applications for TIPS include power plants, plumbing, and heavy construction. In fact, uses for plastic piping systems seem to be limited only by one's imagination.

"Plastics"—the famous one-word line from the 1967 film *The Graduate*—was said to be the key to the future. Never was this prediction truer than it is today for the upcoming revolution of U.S. industry converting wherever possible to TIPS.

Note: Portions of this article were adapted from the book Plastic Piping Systems, 2nd Ed, by David A. Chasis, published by Industrial Press Inc.

LIFE CYCLE ASSESSMENT AND PLASTIC PIPING

More than most industries, the plastics industry uses abbreviations and acronyms galore. For example, acrylonitrile-butadiene-styrene is very rarely stated as such; instead, ABS is commonly used. How about the tongue twister Polychlorotrifluoroethylene when PCTFE would suffice? There is, however, a fairly new acronym that all in our industry should become familiar with: LCA. No, it's not a polymer, but a very important process called Life Cycle Assessment.

Figure 1
Life cycle assessment of plastic pipe.

Plastic critics, usually funded by special interest groups having focused agendas, are attempting to produce a groundswell to limit or ban halogen-containing products such as vinyls and fluoropolymers. The eventual goal of these critics is to completely deselect the use of all plastics from our environment. But are replacements for plastics

really better for the environment? This question is where
LCA can offer a clearer picture.

Life Cycle Assessment is a scientific, unbiased analysis
of a selected product examining the product's total environmental footprint on our planet. The process does not have
any preconceived agenda and follows the similar approach
of Jack Webb's TV persona, Joe Friday, who on the old TV
series *Dragnet* used to say when trying to solve a crime,
"All we want are the facts, ma'am."

It doesn't matter what product is under scrutiny—the
process for a full LCA is the same and is governed by the
requirements of the international standard, ISO-14040.
Generally, Step one of the assessment process determines
the quantity of energy used and the environmental impacts
that occur for the initial stage of processing raw materials to
a final product to be shipped to processors. In the case of
plastic piping, the first part includes the processing of feed
stocks into resin or compound shipments. Step two analyzes
the impacts of processing the received raw material to the
shipment of the finished product. Step three entails the
installation and use phase, which follows the process from
shipment a) to the sales channel, b) to the installer, c) to the
end-user, d) to the life of the product. The last step of the
process analyzes the final disposition of the product after the
product's useful life. The plastic industry's stance is that
plastics and their material competitors should be judged not
in a vacuum but scientifically with proven methodologies
such as LCA.

Attempting to address building construction concerns
and the desire for Green Building ratings, the U.S. Green
Building Council (USGBC), a non-government organization has created a building rating system called LEED
(another acronym standing for Leadership in Energy and
Environmental Design). The USGBC is presently in the

midst of assigning credits to environmentally friendly building features and practices. There is one proposal being presented to the LEED committee to credit the elimination of
all halogenated products from health care construction. If
this proposal was to be adopted by LEED and later carried
over into legislation for health care facilities, there most
likely would be a tendency to limit or prevent the use of
vinyls or fluoropolymers in new health care facilities and, in
the future, other future commercial building projects. This
proposed de-selection credit goes against the findings of the
USGBC's own Technical and Scientific Advisory
Committee (TSAC) report on Polyvinyl Chloride (PVC)
products. The finding of the report was that giving a credit
for the avoidance of a material could result in selection of
products that actually do worse. Obviously, an LCA would
have been a better path forward.

Figure 2 Much more plastic pipe footage can be shipped per truck than other materials.

The Plastic Pipe and Fittings Association (PPFA), the
Vinyl Institute (VI), and the American Chemical Council
(ACC) are vigorously fighting this prescriptive proposal,
which doesn't use any recognized scientific process. One of
the strategies that are being forwarded by the plastics industry is for any green and sustainable building ratings or standards system to use LCA whenever possible to determine
the effectiveness and environmentally soundness of any
building product or material.

Plastics have become the material of choice in many industries due to their durability, ease of installation, and cost-effectiveness. Now using the LCA process, preliminary reports are giving conclusive proof that most thermoplastic piping products—compared to other piping materials—cause less negative impact to people and the environment. The distinct possibility of plastics being generated from renewable feed stocks in the near future is expected to further improve the benefits of thermoplastic piping.

Join the movement to promote products that benefit our world now and in the future...PLASTICS.

Reprinted with permission of the IAPD; issue april/may 2008 – **the IAPD magazine**

LIGHT BULBS AND PIPE

What would you say about two products that save energy galore, last much longer than the products they are replacing, and are cost-effective? Would you conclude that it meets most of the guidelines of being considered environmentally sound? What are these products: compact fluorescent lamps (CFL) and Polyvinyl Chloride (PVC) pipe?

There's hardly a piece of literature that addresses "greenness" issues and doesn't laud the need to replace incandescent lamps (ILs) with CFLs. Why? Because over the life span of a CFL lamp compared to an incandescent lamp, there is a savings in the United States of over $30 of electric costs and savings of over 2000 times the CFL's own weight in greenhouse gases. Plus, the life span of CFLs is 12 to 15 times longer than that of ILs. It is true the purchase price of CFLs are 3 to 10 times the costs of ILs, but taking into consideration the extended life time and low energy use, the CFLs return on investment is extremely reasonable.

Figure 1 Compact fluorescent lamp (CFL).

PVC piping has many similar "greenness" attributes as CFL and, in some areass offers greater environmental benefits. For example, PVC in many applications doesn't have any discernable end-life. That's right; this piping product could last for decades without having the need to be replaced. And unlike CFL, PVC pipe is completely recyclable both in the processing stage (Post Industrial) and the after-life stage (Post Consumer).

The North American piping industry is presently involved with studies called Life Cycle Assessment (LCA) in which scientifically conducted investigations conducted by a third party are determining the economic, energy, and environmental impact of various piping materials. Preliminary findings of the first two phases of the LCA are showing that PVC is in the top echelon of piping materials when it comes to having a light footprint on Mother Earth.

In addition to the differences of prolonged product life and complete recyclability of PVC versus CFL, PVC piping is one of the lowest priced piping offerings based on product, installation, and maintenance costs. CFLs are one of the highest cost lamp bulbs in the market place. But the most significant difference of these energy saving products is how they are disposed after use. Because CFLs use mercury (a rather toxic material) vapor as an ionizer mechanism to produce light, the disposal instructions are similar to other hazardous wastes. The instructions to a consumer of what to do when a CFL breaks include the following:

a. Open a window and vacate the area for 15 minutes.
b. Wearing gloves, carefully clean up the glass and any loose white powder putting all waste in doubled **plastic** bags.
c. When vacuuming over the spillage for the first time, remove the vacuum bag when finished and empty

and wipe the canister putting the bag and vacuum debris as well as the cleaning materials in two sealed **plastic** bags.

There are no hazardous warnings when disposing of PVC pipe. As mentioned before, the piping material is completely recyclable with regrind used in drainage pipe, decks, fencing, and other lightweight, corrosion resistant products. If the pipe isn't recycled and is incinerated or added to landfills, there is no significant harm to the environment. This is not the case with CFL disposal.

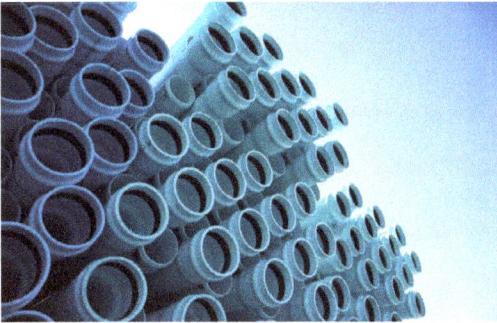

Figure 2
PVC pipe for
underground use.

With this information in hand, why is it that many environmentalists worldwide are vigorously promoting the replacement of incandescent lamps with compact fluorescence lamps while seemingly ignoring the many environmentally proven benefits of PVC replacing failing non-plastic piping systems? Could it be a greenness bias?

CFLs have advantages in illuminating our homes, buildings, and factories while reducing greenhouse gases, but PVC piping does even more—it is one of the most durable, easy and safe to install, environmentally sound, and cost-effective piping systems in existence. It's time for the world to be more aware of the many features and benefits of PVC piping.

Authors note: Since the publication of this article, LED (light-emitting diode) lamps have been introduced; they have become well accepted in the lighting market. LEDs cost more to purchase, but are more efficient and have much greater life spans than either incandescent or fluorescent lamps. In the long run, LEDs are actually more cost effective than other lamps and they do not have significant traces of mercury. However, LEDs do have high lead, arsenic, and heavy metal levels. Future LED technology may reduce toxic material contents, but for the present products, it would be prudent to dispose of LED lamps similarly to that of CFL lamps.

Reprinted with permission of the IAPD; issue december 2009/january 2010 – **the IAPD magazine**

PLASTIC PIPING IN INDUSTRIAL APPLICATIONS

Would you believe that plastic piping has been in use for almost eight decades? It's true. Plastic piping was introduced as an alternative to more traditional, failure-prone piping materials. In pre-plastic days, underground piping came in an assortment of material: metal, concrete, and even wood (yes, in the late 1890s and early 1900s, circular lumber was hollowed out and jointed by metal bracketing). But these all ultimately failed when challenged by corrosion and chemical attack. Through the years, in contrast, plastic piping proved its mettle. Today, for its stellar qualities, it has become one of the leading piping materials for such *underground* applications as water mains, sewer lines, drainage and waste lines, natural gas distribution, swimming pools, sprinkler irrigation, conduit, and trenchless pipe rehabilitation. So why plastic piping?

Figure 1 Corroded metal valve.

There are four major characteristics that account for the success of plastic piping: durability, ease and safety of installation, environmental soundness, and cost-effectiveness. These are the reasons that architects and engineers have been specifying plastic piping over the last 50 years in above ground applications. Plastics picked up where metals failed. Why? Because plastics hold up to the rigors of harsh environments and applications, such as chemical waste drainage, surface finishing (plating), salt water use, ultra and high purity water, mining, water and waste treatment, and chemical processing.

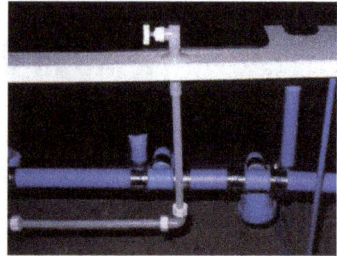

Figure 2 PP chemical waste drainage and high-purity piping.

Even today, as more engineers and installers learn about the benefits of plastic piping systems, they realize that the use of conventional or exotic metal alloy piping is not always the best solution to handle chemically corrosive fluids. Depending on conditions of service, plastic piping systems can handle almost any fluid imaginable. This includes inorganic and organic acids, bases, salts, and slurries. Not only is the plastic piping system chemically resistant, but also the savings in material, installation, and maintenance costs produce unmatched savings for the life of the installation when compared to most other piping materials.

One of the earliest success stories of plastic piping is its use in chemical waste drainage systems. About forty years ago, glass pipe and fittings was the preferred piping material for this application. Sure, glass has excellent chemical resistance. But when you consider its associated product, transportation, and breakage costs, plastic has proven to be a better replacement. Today plastic piping is the market leader in applications of chemical waste drainage systems. Polypropylene (PP), Polyvinylidene Fluoride (PVDF), and Chlorinated Polyvinyl Chloride (CPVC) are used extensively in laboratory waste systems in schools, research centers, and chemical processing plants.

Figure 3 PVC and CPVC piping in a chemical processing plant.

Salt water wreaks havoc on metal piping products. Not so on plastic piping. Depending on conditions of service, Polyvinyl Chloride (PVC) and Polyethylene (PE) are used extensively on salt water environments below 140°F/60°C in such applications as: intake and outtake power plant lines, marine use, fish hatcheries, aquariums, water theme parks, and desalination plants. For handling the field challenges of higher temperature salt water applications, CPVC, PP, Polyethylene-raised temperature (PE-RT), Polyethylene cross-linked (PEX) and PVDF are used.

Nascent industries such as the manufacturing of circuit boards, semiconductors, and other electronic devices have embraced the use of plastic piping. On the one hand, these industries must process very corrosive chemicals in their manufacturing, but also need applications for handling the purest water possible for cleaning silicon chips and boards. There are PVC, PP, and CPVC products to handle the chemical processing applications, and PVDF to handle ultra-high purity water. For these high-purity water applications, PVDF piping, fittings, and valves are cleaned and bagged at the manufacturer's plant for shipment to a semiconductor site. The piping system is then installed in plant clean rooms using very sophisticated joining equipment that minimizes or eliminates possible joint contamination of any kind. It's a given: when a semiconductor plant is on the drawing board, millions of dollars of PVDF and other plastic piping systems will be specified.

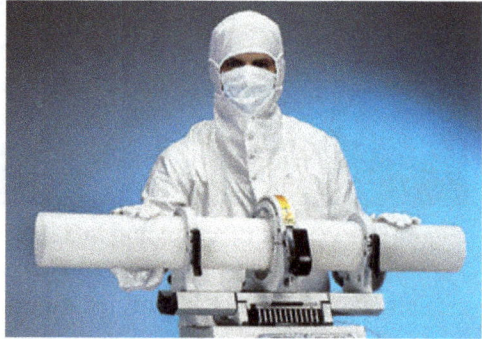

**Figure 4
Joining ultra-high
purity PVDF
piping in a
semiconductor plant.**

Another point to keep in mind with plastic piping is that its remarkable durability does not make for much end-of-life material available for recycling. Yet, all of the plastics mentioned in this article are completely *post-consumer* recyclable. In *post-industrial* recycling (materials recycled in the manufacturing process) more than 99% of all plastic compounds are usually incorporated in the final product.

In the past, the lack of easy access to educational programs and literature has been a factor in keeping engineers from specifying more industrial plastic piping systems. This industry failing has now been remedied. With a simple click, anyone can access excellent websites offering useful information on design, features and benefits, installation, joining techniques, material selection, product availability, and market applications. These websites are:

- Plastic Pipe and Fittings Association (PPFA): www.ppfahome.org

- International Association of Plastic Distribution (IAPD): www.iapd.org

- Sustainable Piping Systems: www.sustainablepipingsystems.com

- Plastics Pipe Institute (PPI): www.plasticpipe.org

- Mechanical Contractors Association of America (OPUS): http://opus.mcerf.org/

The next time you are considering an industrial challenging fluid-handling application—think plastics.

Reprinted with permission of the IAPD; issue February/march 2012 – **the IAPD magazine**

24

PLASTIC PIPING SYSTEMS...
HERE'S TO YOUR HEALTH!

Because of its dramatic success in penetrating markets where traditional piping materials have been used, plastic pipe has become one of the most criticized of any fluid-handling products. Often, the plastic piping industry must coun-

teract the unrelenting attacks by non-plastic rivals, labor unions, and environmental extremists.

To provide proof to architects and specifying engineers, and in response to skeptics, the industry has launched and endorsed Life Cycle Assessment (LCA). LCA *scientifically* analyzes the environmental burdens of a product's life from raw material extraction to the product's end-of-life resolution. The first phase of the analysis, including raw material extraction through pipe manufacture, is completed. Based on the preliminary LCA report findings of North American companies, plastic piping has significantly less environmental burdens than competing piping materials.

Yet, one of the areas not covered in this and other LCA studies is the safety of workers who are involved producing piping raw materials and finished piping products. Fortunately, this safety data has been collected and is available from the United States Bureau of Labor. Safety performance in most industries has improved over the last fourteen years based on OSHA reporting of injuries and illnesses (IIR). Surprisingly, the trend shows over this time period that the plastic piping industry is well below the average IIR rate when compared to other non-piping industries including agriculture, retail trade, and general manufacturing.

Table I lists the 2006 injury and illness rates for selected piping industries based on total U.S. industry reporting of 111,273,100 employment hours with an average incidence rate of 4.4 per 100 fulltime workers. As you can see from this table, the labor force involved in producing plastic piping feed stocks and products have appreciably fewer injuries and illnesses when compared with other piping systems. (Author's note: if concrete and vitrified clay pipe were in the breakdown, plastic pipe would still show fewer injuries and illnesses compare to these piping products).

Next time you meet an anti-plastic piping group that has "green," "health," "justice," or 'peace" in their association name; ask them if it is their intention to promote non-plastic piping materials to the detriment of both the labor force and the environment. Plastic Piping Systems...here's to your health!

Table 1
2006 Incident Rates of Injuries and Illnesses of Selected Piping Industries

Piping Materials	*NAICS No.	**Incident % Plus/Minus RateVS Ave. Rate (4.4)
Iron & Steel		
Iron Ore Mining	212210	2.4
Iron/Steel Mills	331111	5.4
Iron Foundries	331511	15.1
Iron/Steel Pipe & Tube Mfgrs.	331210	9.5
Average Rate	8.1	+ 84.1
Copper		
Copper Ore Mining	212234	3.4
Copper Foundries	331525	8.0
Copper Smelting/Refining	331411	4.4
Copper Roll/Draw/Extruding	331421	10.1
Average Rate	6.5	+ 47.7
Plastic		
Oil/Gas Extraction	211	2.0
Petrochemical Manufacturing	325110	1.0
Petrochemical Refineries	326122	1.4
Plastic Material & Resin Mfgrs.	325211	3.0
Chemical Manufacturers	325000	2.9
Plastic Pipe & Fittings Mfgrs.	326122	7.0
Average Rate	2.9	− 34.1

* NAICS is the code for North American Industry Classification System
** The incidence rate combines injuries and illnesses as reported by the
U.S. Bureau of Labor

Reprinted with permission of PS&D/Plumbing Engineer; issue december/ january 2009 – **the IAPD magazine**

PLASTICS LEARN THEIR ROLE IN PLUMBING SYSTEM DESIGN

Plastics are a suitable piping product in many underground applications such as irrigation, swimming pools, drainage, well casings, sewer lines, slip-lining, hot and cold water distribution, and national gas distribution. This article explores these applications as well as the many aboveground opportunities for plastic pipe usage in industrial and commercial applications. For purposes of this article, soft flexible tubing, plastic-lined metal piping, and fiber-reinforced piping are excluded.

Introduction to Plastic

Plastic is a synthetic material made from polymeric substances of large molecular weight. Although plastic is solid in its finished state, it can be shaped by flow at some stage in its manufacture or processing. The word *plastic* is derived from the Greek word *plastikos* or the Latin word *plasticas,* both meaning "to be molded."

The plastic industry celebrated its 100-year anniversary in 2007. The first all-synthetic plastic, phenol formaldehyde (Bakelite), was discovered in 1907. However, it was not until World War II and post-war industrial demands that polyvinyl chloride (PVC) and polyethylene (PE) plastics were used for piping. Acrylonitrile butadiene styrene (ABS) and polypropylene (PP) were introduced in the 1950s. Chlorinated polyvinyl chloride (CPVC), polyvinylidene fluoride (PVDF), and cross-linked polyethylene (PEX) were introduced in the 1960s.

The plastics used in piping today are compounds formulated from basic resins such as PE, PVC, PP, and CPVC with various additives (see Figure 1). The additives are incorporated in the resin or added during the manufacturing process to provide the desired product performance characteristics.

**Figure 1
Plastic resin
and compounds.**

The two most commonly used manufacturing methods to create plastic piping products are extrusion and injection molding. All thermoplastic piping is extruded. Extrusion is a process whereby heated plastic forced through a shaping orifice becomes one continuous formed piece. Most voluminous thermoplastic non-pipe products are injection molded. Injection molding is the process of forming a material by forcing it, under pressure, from a heated cylinder through a sprue (runner) into the cavity of a closed mold (see Figure 2).

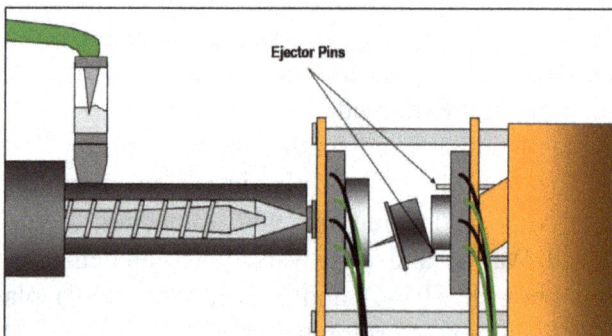

Figure 2 Plastic injection molding process.

Almost all plastic piping used in North America is made from materials having common physical characteristics as classified by ASTM International standards. Despite a movement to redefine the classifications in a more meaningful way, the old designations are still in use. For example, in the past (and also being used presently in piping markings), PVC pipe material commonly had the designation 1120. The first digit depicts the type of PVC, the second digit is the grade of PVC, and the last two digits equal the hydrostatic design stress of the material divided by 100. The new ASTM designation for PVC is based on cell classification, which considers the material and its impact strength, tensile strength, modulus of elasticity, and heat deflection temperature. A typical PVC cell classification is 12454.

To ensure compatibility of plastic piping materials, ASTM has created test methods for determining such physical characteristics as specific gravity, tensile strength, modulus of elasticity, flexural strength, impact strength, coefficient of thermal expansion, thermal conductivity, and others.

Features and Benefits

All thermoplastic piping materials offer the following features and benefits:

Chemical Resistance

This applies to both the interior of plastic piping, which is unaffected by abrasive or otherwise contaminated materials flowing through the piping, and the exterior of the piping, where installation in aggressive soil and water conditions may damage other piping materials. Plastic piping does not require a protective coating in these situations. In fact, plastic piping often is recommended because its chemical resistance leads to longevity.

Corrosion Resistance

Plastic piping is nonconductive and therefore immune to galvanic or electrolytic corrosion attack (see Figure 3). No expensive prevention equipment or coatings are required for underground installations.

Figure 3 Non-corroding plastic spool piece.

Optimum Flow

Because of their smooth interior surfaces, almost all plastic pipes have a Hazen-Williams C factor of 150. Less turbulent flow lowers velocities and lessens horsepower requirements.

Lightweight

Most plastic piping is one-sixth the weight of other piping materials. The light weight dramatically lowers transportation cost, simplifies storage and on-site handling, and allows safer and easier piping installations (see Figure 4).

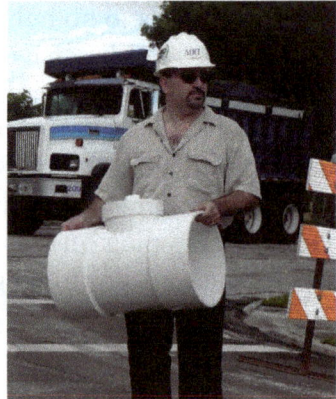

Figure 4 Easily handling a plastic 12-inch diameter tee.

Abrasion Resistance

Many plastics are much more resistant to abrasion than non-plastic pipe, which means less wear on the pipe, especially when slurries are involved.

Low Thermal Conductivity

All plastic piping materials have low thermal conductance, resulting in less heat loss through the wall of the pipe. In many cases, insulation may be eliminated or greatly reduced from the installation.

Variety of Joining Methods

Every plastic piping system has three or more joining options. Two of the joining methods, heat fusion and solvent cementing, create joints stronger than the pipe or fitting being joined.

Variety of Colors

The manufacturing process for plastic piping allows color to be an integral part of the piping system. Color-coding pipes allows a visual safety factor in underground piping applications (see Fig. 5).

Figure 5 Color variety in plastic piping

Nontoxic, Odorless, and Biological Resistance

Plastic piping systems are odorless, nontoxic, and resistant to fungi, bacteria, and termite attacks. These characteristics allow plastics to be used in the food and beverage industry as well as high-purity piping systems. In high-purity and ultra-high-purity water systems, specially formulated non-pigmented PVDF and PP piping materials frequently are used to obtain maximum water purity results.

Code Acceptance

Dozens of plastic piping standards are referenced in plumbing and mechanical codes, including the International Plumbing Code and Uniform Plumbing Code. Some of these standards include:

- ASTM D1785-06: *Standard Specification for Polyvinyl Chloride (PVC) Plastic Pipe, Schedules 40, 80, and 120*

- NSF/ANSI Standard 14: *Plastic Piping System Components and Related Materials*

- UL 94: *Standard for Flammability of Plastic Materials for Parts in Devices and Appliances*

- FM 4910: *Fire-safe Plastics for Semiconductor Equipment*

Integrated Piping Systems

Most plastics come in complete systems of fluid-handling and air-handling products, allowing one plastic material to be in contact with all wetted parts.

Ease of Fabrication

Thermoplastic piping materials can be easily fabricated into many diverse products due to ease of construction.

Ease of Product Identification

Unlike copper and cast iron pipe, most plastic piping products have surface markings showing country of origin, material, pipe size, pressure rating, manufacturer, applicable certification or listing agency, and manufacturing process cycle (see Figure 6).

Figure 6 Typical plastic pipe code markings.

Environmental Soundness

Every thermoplastic piping product material is recycled and used in the manufacturing process with no appreciable material waste and is completely recyclable after the end of product life. Also, the cumulative energy required to manufacture, install, and transport plastic piping systems is much less than that of most non-plastic piping systems.

Cost Comparison of Installed Piping Systems

Most independent studies show that plastic piping systems in industrial, residential, and commercial applications are 20 percent to 50 percent less costly to install and maintain.

Engineering Design Considerations

The accepted engineering practices when designing with thermoplastic piping are similar to those when designing with other piping materials. However, design engineers should be aware of some unique properties to ensure an effective and long-lasting plastic piping installation. These properties are:

- Chemical resistance

- Pipe and system pressure ratings

- Temperature limits

- Temperature/pressure relationship

- Expansion and contraction
- Support spacing

Chemical Resistance

Plastics in general have excellent chemical resistance. This quality is one of the major reasons why plastics have made inroads in many piping applications. Generally, all plastic piping can handle aggressive water systems, including salt water. Most plastic products also handle mild acids and caustics. Piping and resin manufacturers have tested hundreds of reagents to determine their effect on plastic piping. As a result, they have published chemical-resistance tables for design use that list the plastic type plus temperature limitation for each chemical. Figure 7 shows broad general categories of chemical resistance by material.

Chemical - Inorganics	CPVC	PE	PP	PVC	PVDF
Acids, dilute	R	R	R	R	R
Acids, concentrated	R	L	L	R	R
Acids, oxidizing	R	NR	NR	R	R
Alkalis	R	R	R	R	R
Acid gases	R	R	R	R	R
Ammonia gases	NR	R	R	L	R
Halogen gases	L	L	L	L	R
Salts	R	R	R	R	R
Oxidizing salts	R	R	R	R	R

Chemical - Organics	CPVC	PE	PP	PVC	PVDF
Acids	R	R	R	R	R
Acid anhydrides	NR	L	L	R	L
Alcohols-glycols	L	L	R	NR	R
Esters / ketones / ethers	NR	L	L	NR	L
Hydrocarbons – aliphatic	R	L	L	L	R
Hydrocarbons – aromatic	NR	NR	NR	NR	R
Hydrocarbons – halogenated	L	NR	NR	L	R
Natural gas	L	R	R	L	R
Synthetic gas	NR	L	L	NR	R
Oils	L	L	L	L	R

R = Recommended   L = Limited Use   NR = Not Recommended

Figure 7 Typical chemical-resistance table.

Pipe and System Pressure Ratings

There are two basic pressure ratings of plastic piping: schedule and constant pressure. Schedule pipe has an outside diameter similar to iron pipe size, with a wall thickness that matches the wall thickness of the same size as schedule metal pipe. Most vinyl (PVC and CPVC) pipe is available in both Schedule 40 and Schedule 80 dimensions. Schedule pipe pressure ratings vary with pipe diameter. Schedule pipe pressure ratings normally decrease as the pipe diameter increases.

Standard dimension ratio (SDR) pipe is based on iron pipe size outside diameters, with pipe wall thicknesses varying to allow the pipe to have a constant pressure rating for all diameters. Commonly used SDR ratings and corresponding pressure ratings are shown in Table 1. Most metric piping is also constant pressure rated, but instead of pounds per square inch (psi), atmosphere or bar ratings are normally used.

Table 1 Comparisons of SDR PVC pipe pressure ratings at 73°F/23°C

SDR Rating	Pressure Rating (psi)	Bar Rating (atm)
13.5	315	21.4
17.0	250	17.0
21.0	200	13.6
26.0	160	10.9
32.5	125	8.5
41.0	100	6.8

Plastic pipe fittings have similar pressure ratings as the pipe. However, some molded fitting manufacturers have lowered the pressure capability of their fittings in comparison to pipe for critical applications. One of the limiting pressure ratings in plastic piping systems is the 150-psi working pressure rating of valves, unions, and flanges. However, several plastic valve manufacturers now offer valves rated at 235-psi working pressure.

Direct threading of thermoplastic piping is accomplished using only proper threading equipment. Schedule 80 and Schedule 120 pipe are the only plastic piping materials suitable for threading. With vinyl piping, direct threading reduces working pressure by 50 percent. With other materials, the pressure rating is reduced to 20 psi or less. If threaded thermoplastic piping systems must be used, increased working pressures can be obtained using transition fittings such as molded unions and adapters.

Temperature Limits

Although some engineered thermoplastics can handle temperatures exceeding 400°F/204°C, they are not readily available in commonly used piping systems. The maximum temperature limits for commonly used plastic piping are shown in Table 2. The lower limit for plastic pipe is 0°F/–18°C or less. However, when thermoplastic piping will be exposed to temperatures below freezing, the installation must protect the pipe surface from heavy impact to accommodate a decline in the pipe material's shear strength.

Table 2 Maximum temperature limits for plastic pipin

Pipe Type	Temperature Limit (°F) (°C)
PVC	140 60
PE	160 71
PP	180 82
CPVC	210 99
PVDF	285 141

Temperature/Pressure Relationship

Thermoplastic piping materials decrease in tensile strength as temperature increases. This characteristic must be considered when designing plastic piping systems. The correction factor for each temperature and material is calculated by an established formula. To determine the maximum suggested design pressure at a given temperature, multiply the base pressure by the correction factor. Use Tables 3 and 4 to determine the maximum pressure rating of 3-inch PP Schedule 80 pipe at 120°F/49°C. Use the pressure rating of 3-inch PP pipe at ambient, which equals 190 psi, and multiply by 0.75, or 142.5 psi.

The temperature correction factors for valves, unions, and flanges are different than those for pipe (see Table 5).

Table 3 Comparison of Schedule 80 pipe pressure ratings (psi) at 73°F/23°C

Nominal Pipe Size (in)	PVC/CPVC	PE (SDR 11)[a]	PP[b]	PVDF
½	850	160	410	580
¾	690	160	330	470
1	630	160	310	430
1½	470	160	230	320
2	400	160	200	270
3	370	160	190	250
4	320	160	160	220
6	280	160	140	190
8	250	160	N/A	N/A
10	230	160	N/A	N/A
12	230	160	N/A	N/A

[a] PE is not Schedule 80.
[b] Pipe pressure ratings shown are pipe manufacturer's values.

Table 4 Temperature correction factors for piping

Operating Temperature (°F)	CPVC	PE	PP	PVC	PVDF
70	1.00	1.00	1.00	1.00	1.00
80	1.00	0.90	0.97	0.88	0.95
90	0.91	0.84	0.91	0.75	0.87
100	0.82	0.78	0.85	0.62	0.80
110	0.72	0.74	0.80	0.50	0.75
120	0.65	0.63	0.75	0.40	0.68
130	0.57	0.57	0.68	0.30	0.62
140	0.50	0.50	0.65	0.22	0.58
150	0.42	D/O	0.57	NR	0.52
160	0.40	D/O	0.50	NR	0.49
170	0.29	D/O	0.26	NR	0.45
180	0.25	D/O	D/O	NR	0.42
200	0.20	NR	NR	NR	0.36
210	0.15	NR	NR	NR	0.33
220	NR	NR	NR	NR	0.30
240	NR	NR	NR	NR	0.25

D/O = Drainage only
NR = Not recommended

**Table 5 Maximum operating pressure (psi) of valves*,
unions*, and flanges**

Operating Temperature (°F)	CPVC	PP	PVC	PVDF
73-100	150	150	150	150
110	140	140	135	150
120	130	130	110	150
130	120	118	75	150
140	110	105	50	150
150	100	93	NR	140
160	90	80	NR	133
170	80	70	NR	125
180	70	50	NR	115
190	60	NR	NR	106
200	50	NR	NR	97
220	NR	NR	NR	67
240	NR	NR	NR	52

* Valve and union pressure ratings may vary with each manufacturer.
NR = Not recommended

Thermal Expansion and Contraction

Compared to non-plastic piping, thermoplastics have relatively higher coefficients of thermal expansion and contraction. For this reason, it is important to consider thermal elongation and contraction when designing thermoplastic piping systems. Use the following formula and the y values found in Table 6 to calculate the expansion or contraction of plastic pipe.

$$\Delta L = y \frac{(T_1 - T_2)}{10} \times \frac{L}{100}$$

where:

ΔL = Expansion in pipe (inches)
y = Constant factor (inches/10°F/100 feet, see Table 6)
T1 = Maximum temperature (°F)
T2 = Minimum temperature (°F)
L = Length of pipe run (feet)

Table 6 Y factor for plastic pipe types

Material	*y* Factor
PVC	0.360
CPVC	0.456
PP	0.600
PVDF	0.948
PE	1.000

Example

How much expansion will result in 300 feet of PVC pipe installed at 50°F/10°C and operating at 125°F/52°C? Applying the formula above:

$$\Delta L = 0.36 \ \times \ \frac{(125\text{-}50)}{10} \ \times \ \frac{300}{100}$$

Thus, $\Delta L = 8.1$ inches.

Expansion Loops and Offsets

Forces resulting from thermal expansion and contraction can be reduced or eliminated by providing piping offsets, expansion loops, or expansion joints. The preferred method of handling expansion and contraction is to use offset and/or expansion loops. Expansion joints require little space, but are limited in elongation length and can be a maintenance and repair issue. As a rule of thumb, if the total temperature change is greater than 30°F, compensation for thermal expansion should be considered.

Support Spacing

The tensile and compressive strengths of plastic piping are less than those of metal piping. Consequently, plastic piping requires additional pipe support. In addition, as temperature increases, tensile strength decreases, requiring additional support. At very elevated temperatures, continuous support may be required. Tables 7 and 8 list the support spacing of Schedule 80 piping and pipe support spacing corrections for fluids with specific gravities greater than 1.0.

Table 7 Support spacing of Thermoplastic Industrial Piping Systems (TIPS) Schedule 80 pipe (feet)*

Nominal Pipe	CPVC			PP			PVC			PVDF		
	60°F	100°F	140°F	60°F	100°F	140°F	60°F	100°F	140°F	60°F	100°F	140°F
½	5½	5	4½	4	3	2	5	4½	2½	4½	4½	2½
¾	6	5½	4½	4	3	2	5½	4½	2½	4½	4½	3
1	6½	6	5	4½	3	2	6	5	3	5	4¾	3
1½	7	6½	5½	5	3½	2	6½	5½	3½	5½	5	3
2	7½	7	6	5	3½	2	7	6	3½	5½	5¼	3
3	9	8	7	6	4	2½	8	7	4	6½	5¾	4
4	10	9	7½	6	4½	3	9	7½	4½	7¼	6	4
6	11	10	9	6½	5	3	10	9	5	8½	7	5
8	11	11	10	7	5½	3½	11	9½	5½	9½	7½	7

* Listings show spacing (feet) between supports. Pipe is normally in 20-foot lengths. Use continuous support for spacing less than 3 feet.

Table 8 Pipe support spacing corrections with specific gravities greater than 1.*

Specific Gravity	Correction Factor
1.0	1.00
1.1	0.98
1.2	0.96
1.4	0.93
1.6	0.90
2.0	0.85
2.5	0.80

* Above data is for uninsulated lines. For insulated lines, reduce spans to 70 percent of values shown.

Joining and Testing Methods

Plastic and non-plastic installation, testing, and repair techniques have some distinctive differences. However, always use good practices similar to other piping materials. For example:

- Always have a slow-opening valve at the pump discharge.

- Provide for proper air relief and vacuum break at high points.

- Follow specific manufacturer's installation and safety manual

- Train installers handling the piping material for the first time.

- Use appropriate piping joining tools and accessories.

- Ensure that installed piping is as stress-free as possible.

- Eliminate air from the piping system before testing and startup.

Joining Methods

Before joining any plastic piping products, always perform the following:

- Inspect the products to be joined to ensure that no cracks, gouges, warping, or other imperfections are present.

- Make certain the fitting socket and outside pipe diameter fit as described by the manufacturer.

- Cut all pipe squarely, debur all cuts, and bevel where applicable.

- Thoroughly clean all piping products before joining.

- Keep piping products to be joined at similar temperatures when solvent cementing or heat fusing.

- Use appropriate joining methods (see Table 9).

- Be knowledgeable of the manufacturer's installation procedures.

Testing

Test all plastic piping systems hydrostatically prior to full service unless the manufacturer states otherwise. The exceptions are for specially formulated plastic piping systems designed to transport compressed air or gas.

When testing:

- Test all piping before putting it into service.

- Do not test with air/gas unless manufacturer approves.

- Make sure air is removed from lines to be tested.

- Make certain solvent-welded joints are fully cured before testing.

- Test both carrier and containment pipe in dual-containment piping.

- Minimize surge pressures when filling system to be tested.

- Test pressure should be no more than 1½ times the designed maximum system operating pressure or at the rating of the lowest-rated system component.

- Test belowground piping before completely backfilling. (Leave all joints exposed during testing.)

- If testing at high pressures, only the personnel required for the test should be present.

Table 9 Joining methods by plastic piping material

Joining Method	ABS	CPVC	PE	PEX	PP	PVC	PVDF
Flanging	X	X	X	X	X	X	X
Solvent cementing	X	X				X	
Heat fusion			X		X		X
Mechanical—pressure	X	X	X	X	X	X	
Mechanical—drainage		X			X	X	X
Mechanical—push fit		X	X	X		X	X
Mechanical—quick connect					X		X
Mechanical—transitions	X	X	X	X	X	X	X
Direct threading (Schedule 80)		X			X	X	X

Conclusion

For the past four decades, thermoplastics have been one of the fastest growing piping materials in the world. Plastics have been successful due to their environmental soundness, durability, easy and safe installation, and cost-effectiveness. Plumbing engineers should be aware of plastics' unique properties to ensure effective and long-lasting installations.

Reprinted with permission of PS&D/Plumbing Engineer; issue may 2008
– **Plumbing Systems & Design**

THE REAL GREEN PIPING SYSTEM

What piping material can be locally produced, is naturally abundant, and is renewable? Wood pipe! That's right; pipe made directly from wood—preferably bamboo since it can reproduce itself in a ten-year cycle—seemingly offers all the listed benefits.

Wood piping was introduced in North America in geographical areas where lumber was readily available and cheap, such as the Northwestern United States, Canada, northern California, upper New York, and New England. This "wunderkind" piping material was installed in the late 1800s and early 1900s mainly as an alternative for metal piping. Can you possibly imagine a better or "greener" piping material? Shouldn't we resurrect its piping life?

Figure 1 Wood pipe installed decades ago in New England (United States).

We should, if the market would be willing to accept a piping system that leaks, has weakly fabricated or no wood-made fittings, is difficult to join, is attacked by insects and rodents, is not interchangeable with other piping systems, has an irregular internal bore, is flammable, and has

extremely high installation costs. These and other considerations were the reasons the market chose not to use this
"green" piping system decades ago. The same rationale still
exists.

The reasons any product (pipe or any other item) should
be selected for use is a gestalt approach. Certainly impact on
the environment should be a consideration, but what about
other features such as durability, ease of use, safety, and
cost-effectiveness? Incorporating all of these benefits have
made plastic piping systems one of the fastest growing piping system in the world for the last five decades; in the engineering community, many engineers opine that plastic is the
preferred piping material in many applications.

**Figure 2 Plastic pipe is
durable, lightweight, and
easy to install.**

The plastic piping industry produces products that have
proven themselves year-after-year. It is the most scrutinized
piping product ever made; yet, the plastics industry has to
constantly maintain a defensive posture based on attacks
from environmental and labor groups who have their own
agenda. The shame of the matter is the attacks are for the
most part without scientific proof or merit. If these extremists were truly interested in the needs of community, environment, and society, they would embrace products that
offer optimum benefits to the public.

It's interesting to note that the plastic piping industry has
worked closely with the governmental watchdog agencies

(EPA, FDA, OSHA) for years, as well as dozens of local, national and international code and standard agencies. When any of these agencies have asked for input, tests, studies, or plant visits, the industry has been eager to assist. Not only that, but companies producing plastic piping products for potable water applications pay to empower an impartial third party, such as the National Sanitary Foundation International, to oversee and ensure all ASTM and other standards are being achieved. In addition these certifiers have the authority to close down manufacturing production if these rigid standards and not being met.

Figure 3 Plastic piping is preferred in many drainage applications.

And yet certain groups feel there is a conspiracy afoot in which governmental and third party agencies are not astute enough to determine how plastic piping may cause harm to the environment. These same groups have tried for 20 years or more to attack plastic piping in Europe with little measurable success.

Figure 4 Plastic piping can be coiled to 2-inch diameters and larger.

Presently, the plastic piping industry is having their products evaluated by third-parties to determine the overall scientific impact to the environment and economy of plastic pipe versus other piping materials; these parties are using an analytical tool called Life Cycle Assessment (LCA). This detailed analysis studies products from phases of inception to the end of product life. LCA will be used by green building rating systems and standards as a way to compare fairly the true impacts of products without biased rhetoric. Results in Europe and preliminary results in North America are showing that plastics are the equal or better than most other piping material contrary to extremist's press-released myths.

The next time you are selecting which product is most "green," consider using common sense and proven science to assist your decision-making process—it might prevent you from accepting "wooden nickels."

Reprinted with permission of the IAPD; issue february/march 2008 – **the IAPD magazine**

27

REDUCE GREENHOUSE GAS EMISSIONS
WITH PLASTIC PIPING

We have seen tremendous progress throughout the world in the application of above and below ground plastic piping systems. Again and again, plastic piping is one of the materials of choice for municipal water mains, drain-waste-vent lines, irrigation, chemical waste drainage, swimming pools, hot and cold water distribution, natural gas distribution, and several other markets. Why? It's simple! There are four good reasons. By focusing on these four reasons, we'll see that plastic piping is without equal in reducing greenhouse gases in the environment.

Plastics reduce greenhouse gases because they are:

- Durable

- Easy and safe to install

- Environmentally sound

- Cost-effective

Figure 1 Lighter weight plastic pipe reduces transportation costs.

Durability means a long working life. Several independent research institutions have concluded that the working life of plastic piping is unmatched when compared to piping systems of other materials. How does the durability of plastic piping reduce greenhouse gases? There are fewer pipe failures and less maintenance. The reduced occurrence of pipe failure and reduced requirement for system maintenance mean a reduced need for gas- and electric-powered repair equipment. Also, because plastic piping is so durable, replacement piping doesn't have to be ordered and shipped. Transportation costs are always a consideration. With plastic piping, transportation costs can be reduced significantly, as we'll see later in this article.

Plastics are easier and safer to install than any other piping material. Plastic piping is famous for its light weight. This feature greatly reduces the need for costly lifting equipment. Less equipment leads to fewer emissions released into the environment. Another advantage of plastic's light weight is the likelihood of fewer injuries in the field. In most cases, the tools and techniques for joining plastic piping are inexpensive, easy to use, and provide joint integrity.

Plastic piping delivers on environmental soundness. Because plastics have a high degree of chemical and corrosion resistance, this eliminates the need for protective interior and exterior coatings or liners that other piping materials require. There's more: plastics are highly efficient for

Figure 2 Plastic pipe is preferred in handling gray water applications.

facilitating thermal conductivity and smooth flow. These qualities are important to prevent temperature loss of transported fluids. Still more: plastics, compared to other piping materials, have less friction loss in the smooth inner pipe walls. This important characteristic minimizes the pumping horsepower required to move fluid from point A to point B. What all of this adds up to is energy savings. Less energy produced and used—fewer emissions. Less material manufactured and transported—fewer emissions.

For cost-effectiveness, plastics keep on giving. In almost every application, plastics are the low cost installed piping material. Plastics mean savings in cost of materials, and savings on installation labor, maintenance, and job site thefts. Plastic piping is king of the hill. But there is one more significant cost-saving and environmentally-sound feature that can't be overlooked when considering a piping project: transportation of materials.

Figure 3 Septic systems use plastic pipe to reduce use of main sewage treatment facilities.

Tables 1 and 2 list the weights of common metal and plastic piping materials used in residential, commercial, and industrial construction. The data show that plastic piping is hands down the leader in lightweight piping materials.

So how does this translate to money savings and reducing greenhouse gases? This can best be explained by example. Let's say a commercial construction project requires 30,000 feet of 3-inch diameter piping. Most pipes are pal-

Table 1 Weights[1] of Piping in Drainage and Pressure Applications

		Metal		
Nominal Pipe Size[2]	Aluminum (40)	Copper (L)	Steel (40)	Cast Iron (No–Hub)
1 1/2	0.94	1.14	2.72	2.90
2	1.26	1.75	3.65	3.80
3	2.62	3.33	7.58	5.40
4	3.73	5.38	10.79	7.10

			Plastic			
Nominal Pipe Size[2]	ABS Cell-Core	CPVC (40)	PVC (40)	PVC- Cell-Core	PP (40)	PVDF (40)
1 1/2	0.27	0.60	0.54	0.34	0.40	0.88
2	0.37	0.80	0.72	0.46	0.60	1.25
3	0.75	1.66	1.49	0.90	1.00	1.80
4	1.07	2.36	2.12	1.24	1.50	2.80

Notes: 1. All weights are in pounds per foot
  2. Pipe diameter size in inches

letized and transported by trucks on a 40-foot long flatbed that can carry a maximum load of 40,000 pounds. If we selected the lightest weight plastic piping, ABS Cell-Core, the total pipe shipping weight would be 21,900 pounds. The result would be that the entire 30,000 feet of ABS Cell-Core pipe would be transported in one truck (this would be true of PVC Cell-Core piping also). In many cases the limiting factor in transporting plastic piping is a volume issue. There are only so many pallets of pipe that can be vertically stacked before the heights of the pallets exceed safe highway requirements.

For comparison, what are the numbers for 30,000 feet of 3-inch diameter cast iron no-hub piping? The costs to the buyer and the environment would be significant. The palletized metal piping in this scenario would be shipped in 4 trucks, with each truck carrying 7,200 feet of piping at a total weight of 38,880 pounds. That's right, 3 more trucks would be needed for metal piping with all the associated material and labor costs, plus the additional amount of greenhouse gases that would be emitted to the environment. Keep in mind that in most instances the limiting factor in shipping metal piping is the weight limitation of the transporting vehicle.

Table 2 Weight Comparisons of 3-inch Diameter Piping

Material	Lbs./Ft.	% of PVC 40 Index
ABS (Cell-Core)	0.75	0.50
PVC (Cell-Core)	0.90	0.60
Polypropylene (40)	1.00	0.67
PVC (40)	**1.49**	**1.00**
CPVC (40)	1.66	1.11
Polyvinylidene Fluoride (40)	1.80	1.21
Aluminum (40)	2.62	1.76
Copper (L)	3.33	2.23
Cast Iron (No Hub)	5.40	3.62
Steel (40)	7.58	5.09

To all you specifying engineers and installers, when deciding which piping material is best suited to your client's needs as well as being a great choice for the environment— think plastics.

Reprinted with permission of the IAPD; issue april/may 2012 – **the IAPD magazine**

28

SIMILARITIES OF PLASTIC TAPE AND PIPE

A well known midwest company is celebrating its vinyl electrical tape's 60th anniversary. Like plastic pipe, vinyl (polyvinyl chloride) tape was created to solve problems. The dominant insulating tape in use in the 1940s was composed of tar, cotton, and rubber. This tape was expensive, prone to rotting, not very strong, and difficult to apply. Doesn't this problem scenario sound familiar? In the 1940s, metal piping was expensive, was not easy and safe to join, lacked durability (especially in chemically aggressive environments), and was not very environmentally friendly (although in that era the environment wasn't even a blip on the radar screen of any special interest group).

Figure 1 Rolls of vinyl (PVC) tape.

In a short period of time, vinyl tape—due to its features of strength, stretchability, adhesiveness, and price competitiveness—became the market leader and preferred supplier, not only of the insulating tape market, but also for markets never even dreamed of. And so it is with plastic piping. In a relatively short time period, thermoplastic piping became the preferred supplier in such diverse markets as hot- and cold-water distribution, water mains, drain-waste-vent, sewer, irrigation systems, natural gas distribution, swimming pools, acid/chemical drainage, and several others. Why? Plastics are environmentally sound, easy and safe to install, reliable, long-lasting, and cost-effective.

Another similarity of "wunderkind" tape and pipe is the use of Japan's very valuable contribution in the engineering world: *kaizen*. In English, Kaizen translates into a focused, constant-improvement mentality used in the manufacturing and product design process. Since the introduction of vinyl tape in 1947, there has been a minimum of 16 improvements in this product. Plastic piping manufacturers in the last several decades have made even more improvements than its tape cousins.

Consider some of these innovative piping product improvements in just the last 30 years: high impact and high-temperature compounds, electro-fusion joining methods, translucent and multi-colored compounds, spirally wound drainage piping, ultra-high purity compounds, flame and smoke retardant compounds, reduced volatile, organic compounds (VOCs) used in cementing, joining using infra-red heat fusion, molecular bi-axially oriented pipe compounds, compounds especially designed for compressed air and gases, cross-linked polyolefin compounds, composite piping systems of plastic and metal, liners designed specifically for slipping into failing non-plastic piping systems, spline and grooved couplings for mining applications, injected molded fittings for pressure applications up to 12-inch in diameter, fluoropolymer compounds for chemically challenging environments and temperatures approaching 300°F/149°C, specially designed compounds for fire sprinkler systems, push-fit joining systems for small diameter piping systems, and dozens more.

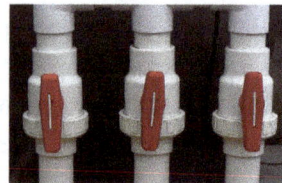

Figure 2 PVC pipe and valves.

With the advent of applied nanotechnology and more advanced organic chemical research, the future of plastic piping has few limits. For many plumbing and mechanical design engineers and installers, change is not easy to accept, especially in the field of piping. But as the market place embraces and rewards new piping technology that provides products that are safe, durable, sustainable, and cost-effective, plastics will continue to lead the way. Just as vinyl tape has become a tool-box stable, plastics will continue to be one of the preferred materials in most piping applications now and in the future.

Reprinted with permission of the IAPD; issue august/september 2007 – **the IAPD magazine**

29

SUSTAINABILITY AND PLASTIC PIPE

The United Nations defines *sustainability development* simply as that which "meets the needs of the present without compromising the ability of future generations to meet their own needs." Let's see how this applies to thermoplastic piping.

Most plastics are created from feedstocks of hydrocarbons extracted from oil and gas. The key word in the last sentence is *hydrocarbons*. In today's world, oil and gas are the least costly hydrocarbon resource to produce feedstocks for plastics. Typically, less than 4% of a barrel of oil (about a gallon and a half) is used for plastic production. But what happens to plastic production when oil and gas reserves dwindle or become extinct? The switch is made to another hydrocarbon resource—coal (it is estimated that the USA has coal reserves that will last for another 245 years). Okay, what happens when coal reserves are depleted?

Plants! That's right, plants provide hydrocarbons for bio-fuels that are currently being used for some plastic production; they will be a renewable source for plastics in the future. In the case of one of the most popular piping materials in the world, PVC, 57% of this polymer's weight is chlorine, which is created from salt, one of the most voluminous materials on earth. The conclusion is obvious: compared to most other common piping materials, plastics have the most sustainable production reserves today and for the foreseeable future.

Figure 1 Various bio-feedstocks to make ethylene.

Having a sustainable resin base is just one part of the equation. What about the economic, energy use, and environmental impact of plastic piping? The good news is these impacts are measurable, beyond the complaints of those activists whose agenda excludes current scientific practices. *Life Cycle Assessment* (LCA) is the process used to determine the economic, environmental, and energy impact of a particular product. By analyzing each piping material phase from raw materials—to resin production—to pipe manufac-

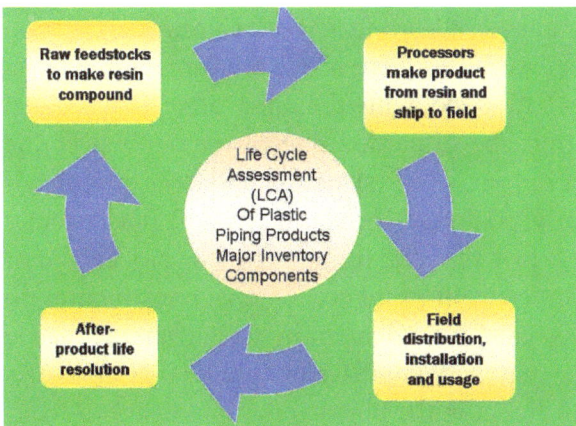

Figure 2
LCA chart for
plastic piping.

**Figure 3
Plastic piping is
extremely light in
weight.**

ture—to field use—to final distribution, conclusions can be reached determining which piping material offers the end-user the best choice. The international piping industry, both plastic and non-plastic, has conducted and is continuing to conduct life cycle studies of their products. So far, completed studies and preliminary results of present studies puts plastic piping in a very favorable light compared to other piping materials.

Does plastic piping meet the above definition of sustainability? Absolutely! The material is light in weight, resistant to many chemical and corrosive environments, safe and easy to install, durable, and cost-effective.

What about recycling? In the process of manufacturing plastic piping, almost all product wastes in processing can be reground and reused. If there is one concern about plastic piping, it is what to do with the pipe after it has been installed and must be discarded due to remodeling or the demolition of an existing structure. Plastic piping, if installed properly, could last for many decades. It's basically inert. Another fact is that due to the very low percentage of plastic piping in residential and commercial construction (in many cases less that ½ of 1 percent by weight), most plastic piping today is put into landfills or incinerated. There is presently a movement in the plastic piping industry to rectify this situation and have manufacturers take back and recycle their product, assuming it can be done to code standards and in a cost-effective manner for the owner and supplier.

Sustainability touches us all in practically everything we do—whether it is the car we drive, the house we live in, the vacations we take, the products we buy, or the hobbies that relax us. No matter what we do, we need to leave a better tomorrow for the next generation. In this regard, the use of plastic piping will continue now and in the future to provide a positive impact on our planet.

Reprinted with permission of the IAPD; issue april/may 2010 – **the IAPD magazine**

30

SUSTAINABLE PIPING SYSTEMS
FOR GREEN BUILDING

Plastic is the most widely used piping material in the world. It is extremely durable, easy and safe to install, environmentally sound, and cost-effective. These attributes have made plastic the preferred piping material in such markets as chemical waste drainage, aquariums, hot and cold-water distribution, irrigation, natural gas distribution, residential drain-waste-vent, sewers, swimming pools, and water mains.

In the last three decades, we have witnessed an explosion of innovation in environmentally responsible and resource-efficient technologies. This greening of our outlook is a trend that is driving change—notably, a change in the way buildings are designed and constructed. Widespread adoption of green technologies will have an impact. It will bring improvement to our quality of life and health, reduce our energy consumption, and conserve and protect our precious water resources.

As new green standards are established and must comply with more mandated green building regulations for new construction, plastic piping will continue to play a central role in beneficial applications within the green building industry. For its cost-effective, ease of installation, durability and environmental sound qualities, plastic piping is part of the green solution.

As we examine how plastic piping is used—specifically in buildings that are more sustainable and energy efficient than our traditionally built ones—we will offer a brief description of each green technology, followed by the top five benefits and a summary of plastic pipe usage. For more detailed information on plastic piping materials, please go to the Appendix.

You might be surprised at the myriad ways that plastic piping assist buildings in becoming greener. Plastics are helping to build a better world, benefitting all with:

- Better quality of life and health
- Energy savings
- Water conservation

Figure 1 Plastic piping achieves green standards.

Life/Health Quality

How does one put a value on lives saved or the prevention of long- and short-term disease? This question is difficult to answer. Yet, if green technology means protecting the environment and our well-being, what could be greener, for example, than residential fire sprinkler systems?

Figure 2 Plastic fire sprinkler systems for homes reduce the risk of catastrophic fires.

We'll examine major green technologies that reduce or eliminate harmful conditions that could cause long- and short-term health problems or death:

- Central vacuum
- Double containment
- Foundation drainage
- Radon venting
- Residential fire sprinkler

Central Vacuum

EPA studies show that most people spend 90% of their time indoors and that indoor pollutant levels can be 2 to 5 times higher than outdoor levels. Hence, improving the IEQ (indoor environmental quality) is very important for our well-being, especially for those who may be challenged with respiratory problems. Central vacuum systems have proven to be successful in reducing indoor pollutants and allergens.

The major components of a typical central vacuum system are: stationary air pump, power unit and dirt collection

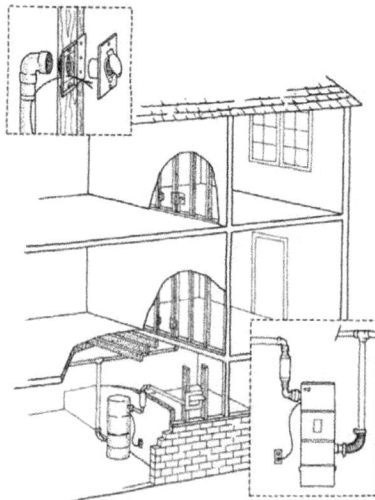

Figure 3
Typical piping
system in home.

Source: Don Vandervort's Hometips.com

canister. The canister is normally installed in a garage, basement, or attic; vented outdoors; and sized to require emptying just a few times a year. Suction vents have hidden connected piping to the canister and are located throughout the home. When a vacuum hose end-connector is inserted into the suction vent, the air pump automatically is activated and starts sucking particulate matter directly into the canister through the vacuum hose and rigid piping. Detaching the hose to the suction vent automatically shuts off the air pump.

Benefits of a central vacuum
- Cleaner indoor air since pollutants and allergens can be removed without releasing them back into the building, as do conventional vacuum cleaners (CVC)
- Two to five times more suction power than CVC
- Large dirt storage capacity requires emptying only once or twice per year
- Not as noisy as CVC
- No need to drag along a CVC throughout the house

Plastic piping for central vacuum systems
Most central vacuums use two-inch diameter PVC piping, although ABS may also be a suitable alternate especially when using a Schedule 40 piping system. Many residential homes use a thin-walled rigid PVC piping system— about half the wall thickness of Schedule 40. Most building codes allow either a thin walled or schedule 40 piping system.

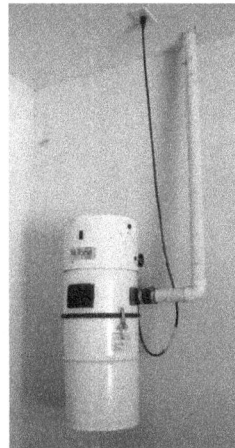

Figure 4 PVC piping attached to central vacuum installed in garage.

Double Containment (DC)

Piping failures in systems that carry harmful fluids can cause contamination of water supplies, as well as atmosphere and ground pollution, and also extensive property damage and/or bodily harm. If piping leaks could inflict catastrophic damage, the design and installation of DC piping must be a prime consideration.

DC piping systems have three basic components: carrier pipe, containment pipe, and spacers or centralizers that keep the containment pipe concentrically supported within the carrier pipe.

The space between the carrier and containment piping is the interstitial or annular space. Any leakage will form in the annular space where leak detection devices are installed in order to set off a remote alarm and quickly pinpoint the leak. DC systems conveniently come as ready to ship and pre-assembled, custom-fabricated or retrofitted to existing piping systems.

Figure 5 PVC X stainless steel DC piping system.

Benefits of double containment piping

- Prevents contamination of groundwater and other freshwater sources

- Prevents harmful fluids from polluting the environment and harming personnel

- Facilitates detection and repair of leaks in carrier piping

- Carrier and containing piping can be of mixed materials

- May reduce insurance premiums by preventing catastrophic spills

Plastic piping for double containment piping

There is practically no limit to the combinations of carrier and containment piping materials that can be used in double containment systems: metal with metal, plastic with plastic, or plastic with metal. Chemical compatibility, material costs, and conditions of service will dictate which piping materials are appropriate.

Figure 6 Clear PVC retrofit DC piping system.

Figure 7 Prefabricated PP DC piping system.

Foundation Drainage

Foundation drainage is designed to divert or direct water away from a foundation. It is installed at the outside foundation border or under the foundation. The purpose of foundation drainage is to prevent problems to the building, foundation, and surrounding soil that excess water causes.

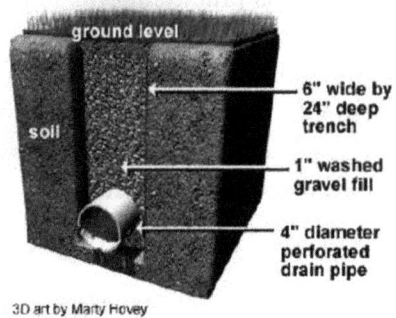

**Figure 8
Typical French drain
design.**

The typical foundation drainage system has perforated piping abutting the foundation/footings with some type of cloth screening material laid on top of the pipe. The pipe and screen assembly is then covered with small diameter gravel and finally covered with a layer of soil or sod. The perforated piping allows for excessive groundwater to percolate into the piping and be drained away to a desired collection area.

**Figure 9 Lengths of PVC
perforated pipe.**

Benefits of foundation drainage
- Keeps excessive water from damaging building foundation
- Reduces or eliminates mildew and mold development
- Can prevent foundation cracking
- Helps prevent damage to landscaping
- Can provide a source of gray water reuse

Plastic piping for foundation drainage

Thinner-walled perforated and non-perforated PVC pipe is used for this application, with DWV Schedule 40 fittings. ABS piping may also be used.

Radon Venting

Odorless, colorless, and tasteless, radon is a radioactive gas that occurs naturally from the decay of underground radium deposits. It is the number one cause of lung cancer among non-smokers. In certain areas of the country, radon can reach dangerous concentrations. It is released into buildings through cracks or holes in the foundation. The EPA states that 1 out of 15 homes in the United States have elevated radon concentrations. Fortunately, there are several radon-venting techniques that can be used depending on the soil and foundation conditions.

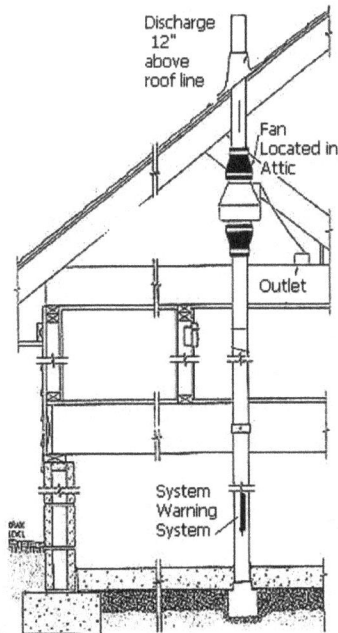

Discharge 12" above roof line

Fan Located in Attic

Outlet

System Warning System

Figure 10 Typical radon venting piping system.

Benefits of radon venting

- Reduces harmful radon concentrations to acceptable levels

- Assists in meeting local, state, and federal radon emission codes

- Reduces the risk of lung cancer

- Minimizes entry of moisture and other harmful soil gasses

- Adds to resale value of house

Figure 11 PVC radon venting piping in home.

Plastic piping for radon venting

PVC Schedule 40 DWV pipe and fittings are mostly used. ABS DWV piping would be an acceptable alternative.

Residential Fire Sprinkler

Fire sprinkler systems in homes save lives, property, and water resources. According to the National Fire Protection Association, 84% of all civilian fire deaths occur in homes. The Home Fire Sprinkler Coalition estimates that smoke alarms and fire sprinkler systems reduce home deaths by 82%. Also, less water is used in fire containment compared to fire hose use. Additionally, there is much less property damage, because 90% of home fires are contained by the operation of just one sprinkler head.

There are two types of residential fire sprinkler systems: stand-alone and multipurpose. The stand-alone system has a separate piping system and water source. Multipurpose systems are combined with an existing cold-water plumbing line.

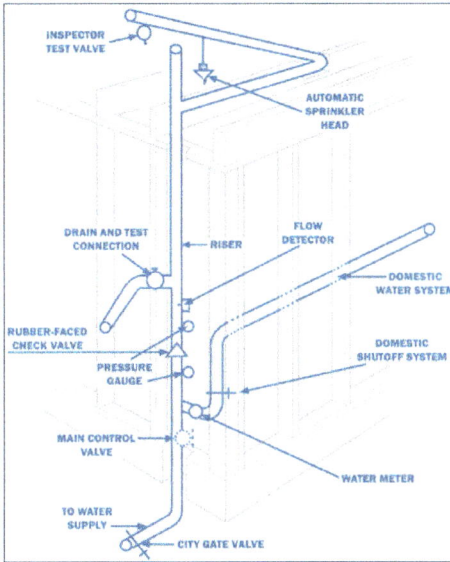

Figure 12
Typical fire sprinkler residential piping diagram.

Benefits of residential fire sprinkler systems

- Save lives and reduce water usage and damage

- Along with smoke alarms, reduce death in homes

- Save 8.5 times the amount of water compared to fire-hose usage

- Usually reduce insurance premiums

- Have proven reliability

Plastic piping for residential fire sprinkler systems

Specially designed CPVC piping has been used as fire sprinkler piping material for decades. More recently, PEX was introduced as an alternative material for this application. CPVC systems are designed to handle a working pressure of 175 psi @150°F/66°C and PEX systems are designed to handle a maximum working pressure of 180 psi at a temperature of 100°F/38°C. CPVC and PEX fire sprinkler systems are approved by most code bodies and agencies.

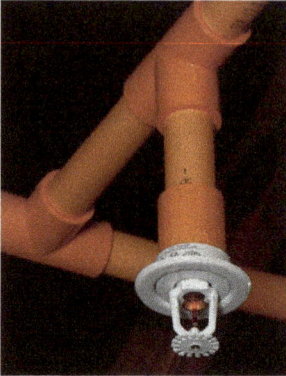

Figure 13 CPVC fire sprinkler plug

Figure 14 PEX fire sprinkler plug

Energy Savings

Buildings in the United States are energy hogs. According to the United States Green Building Council, buildings consume a staggering 72% of the nation's available electric energy. As if that weren't bad enough, this energy use is responsible for 38% of greenhouse gas emissions in America. Any way you look at it, it's a big, unwanted footprint on the environment. To reduce it, we must have greater energy efficiency in our building systems. The goal must be a lowering of energy requirements and greenhouse gas emissions. Such a goal will deliver additional benefits that are no less important: less need to enlarge or build municipal utilities and less fossil fuel usage. Let's look into the following technologies to see how energy can be saved:

Figure 15
Savings reduce power plant usage.

- Decentralized Wastewater Treatment

- Drain-Wastewater Heat Recovery

- Geothermal Energy

- High Efficiency Hot Water Distribution

- Radiant Heating

- Solar Water Heating

Decentralized Wastewater Treatment

Forgoing municipal sewer lines can provide decentralized or on-site wastewater treatment for an individual home or small groups of homes. Decentralized wastewater treatment is constructed at or near the site. It is not connected to a municipal sewerage system. The wastewater is normally returned to the soil.

The major components of a typical decentralized system are a septic tank, plastic pipe, and a permeable soil absorption field for final treatment and dispersal. There are even systems that funnel wastewater to further treatment, resulting in gray water for landscape use.

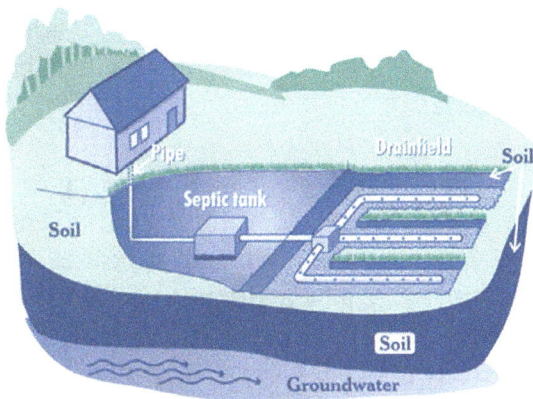

Figure 16 Conventional wastewater home system.

Figure 17
Wastewater system being
treated on site
not being soil absorbed

Benefits of decentralized wastewater treatment

- Disposes wastewater safely, protecting public health and the environment

- Minimizes illegal discharges

- Houses in remote areas can handle wastewater safely and cost effectively

- Conserves the capacity of central treatment plants

- Homeowners generally pay less for decentralized treatment systems

Plastic piping for decentralized wastewater treatment

PVC piping is the preferred material for this application, although ABS and PE drainage piping could be used as alternate materials.

Drain-Wastewater Heat Recovery

Typically 80–90% of the energy used to heat water in homes goes down the drain after use. A heat exchanger and storage tank can make a huge difference. These two devices maximize drain-wastewater heat recovery by re-capturing heat from already used hot water to preheat cold water entering the water heater. A heat exchanger can recover heat from hot water used in showers, bathtubs, sinks, dishwashers, and clothes washers.

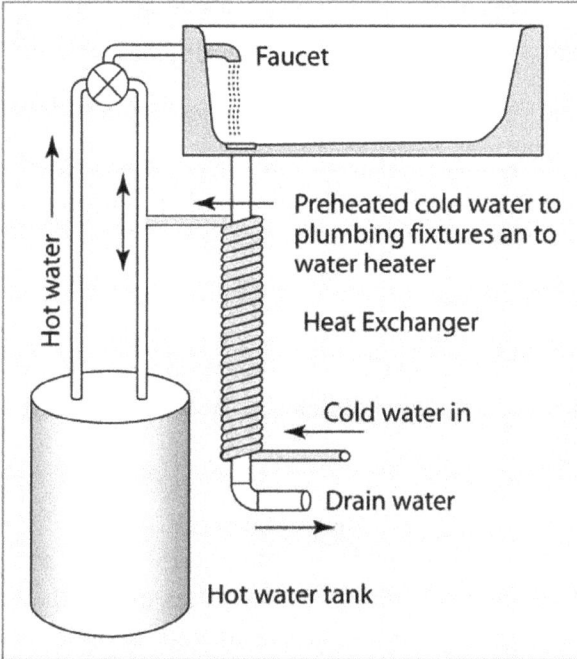

Figure 18 Typical drain wastewater heat recovery.

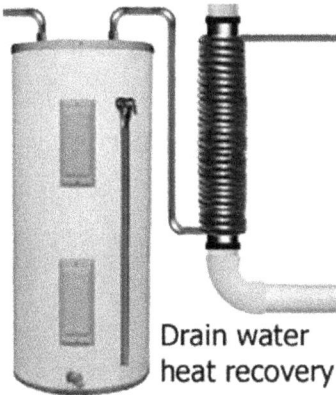

Figure 19 Heat exchanger used in heat recovery process.

Benefits of drain-wastewater heat recovery
 • Saves energy by preheating cold water in water heater and water fixtures

- Increases existing water heating capacity

- Most systems cost less than $700 to install

- More cost-effective if installed in new home construction

- Energy savings offer an excellent payback recovery system

Plastic piping for drain-wastewater heat recovery

CPVC, PEX, PE-RT, PP and composite piping can be used in handling cold, preheated, and hot water processes involved.

Geothermal Energy

A Ground Source Heart Pump (GSHP) is a common geothermal energy technology. It makes use of the constant year-round ground temperatures at depths of less than 10 feet, for example, in most temperate regions. Employing hundreds of feet of buried, liquid-filled piping, this process has the flexibility to either gain or disperse thermal energy, serving as a heat source in the winter and as a heat sink in the summer.

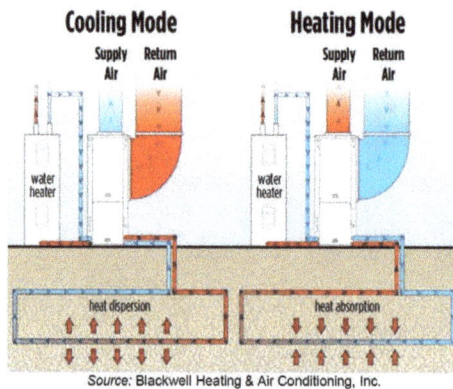

Figure 20 Geothermal heat pump process.

The major components of the GSHP: indoor heat pump equipment, an in-ground piping loop, and a flow center to connect the pump and the loop. A heat-exchange fluid is pumped through the in-ground piping loop, either absorbing heat from the ground in the winter or dispersing heat to the ground in the summer. During cold months, the fluid is circulated through a heat pump, transferring gained ground heat to a forced air or radiant heating system. To disperse heat during hot months, the fluid can also extract heat from the building via the in-ground piping loops, most of which are closed-loop horizontal or vertical in design, depending on the building site.

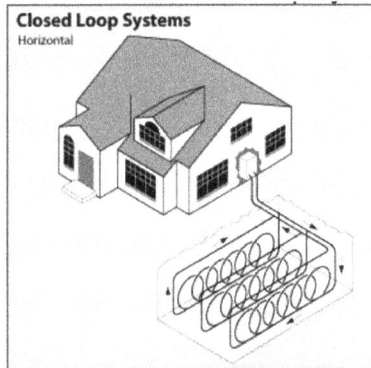

Closed Loop Systems
Horizontal

**Figure 21
Horizontal closed-loop system.**

In additional to ground source thermal energy, there is another geothermal energy system referred to as Enhanced Geothermal System (EGS). This system is used to fuel large power plants by drilling wells vertically 3 to 8 miles into the earth's crust and then utilizing the super-heated water found at that depth to produce steam to drive conventional turbines that produce electricity.

Benefits of geothermal energy
- More efficient than electric-resistance heating, gas and oil-fired systems
- Less expensive to operate and maintain than other systems

- Humidifiers not required since air is not dry

- Reduces greenhouse gases since no carbon dioxide is produced

- Estimates are that 70% of system's energy is from renewable earth energy

Plastic piping for geothermal energy

PE and PE-RT are the major piping players in this technology. PEX and PP may be used in some residential geothermal close-and open-loop systems.

Figure 22 PE closed-loop piping system installation.

Source: Plastic Pipe and Fittings Association

Figure 23 PE loops for geothermal energy systems.

High Efficiency Hot Water Distribution

Efficiently designed hot water distribution reduces the travel time it takes for hot water to be delivered to an appliance or fixture. High efficiency distribution normally includes insulated piping that has as short a run as possible from the hot water heater to the fixture/appliance, with piping sized in as small of a diameter as codes will allow. Also, properly sized, energy-efficient water heaters should be located as close as possible to the end fixtures/appliances.

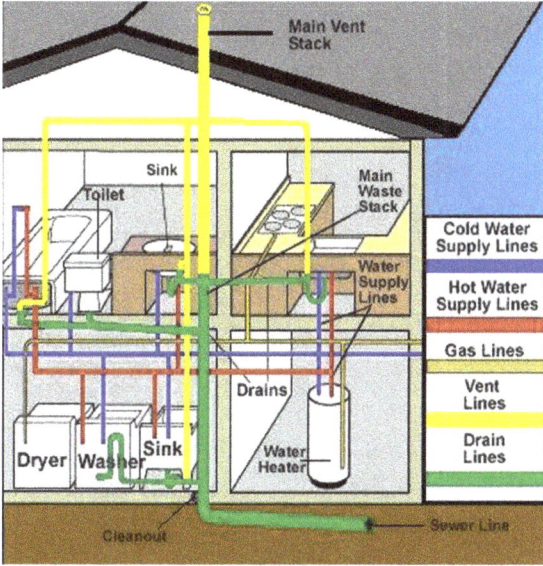

Figure 24 Typical residential piping plan.

There are several major types of recirculation pumping systems: continuous; demand-controlled; temperature-controlled; time-controlled; and time-and temperature-controlled. Demand-controlled pumping is the most energy efficient because the recirculation pump typically runs as little as ten minutes a day, while providing wait times similar to continuous recirculation systems.

An alternative method of achieving high efficiency hot water piping is parallel piping, also referred to as a "home-run" system. Here, a central manifold is used to deliver water to individual fixtures through small diameter piping (usually 3/8 CTS). These lines run directly from the manifold to the fixtures with pipes holding less water volume; therefore, the wait times and water wasted during delivery are significantly reduced. This system is gaining in use since it is less costly to install and operate than most conventional piping systems.

Benefits of high efficiency hot water distribution
- Reduced energy and water utility bills for owners
- Reduced wait times of hot water delivery
- Reduction in wasted energy when water sitting in piping cools down
- Lower installation costs than conventional systems
- Fewer burdens on energy, water and wastewater utilities

Plastic piping for high efficiency hot water distribution
PEX and CPVC are the most commonly used piping materials for this application. For parallel piping systems, PEX is generally the material of choice. Composite piping, PP, and PE-RT are also viable materials for consideration.

**Figure 25
PEX "home run" hot-
and cold-water
distribution piping.**

**Figure 26
CPVC hot-and
cold-water
distribution piping.**

Radiant Heating

This technology heats space by applying heat underneath or within floors, walls, ceilings, beams, or panels of a building. The conditioned space is heated by radiation. Space heating also occurs with natural convective circulation by air rising from the floor.

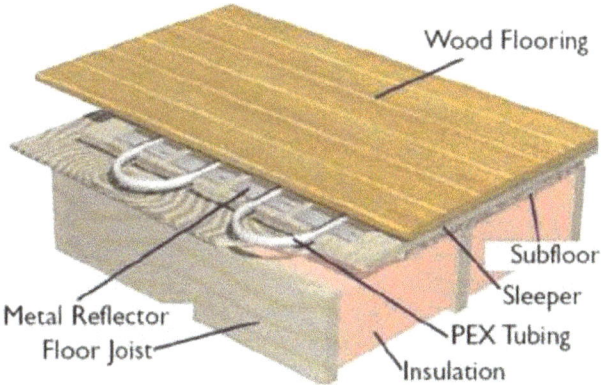

Figure 27 Typical under floor radiant heating system.

There are three types of radiant floor heating systems: hydronic, electric, and air. Hydronic is the most popular. The basic components of the hydronic system are a heat source (boiler, water heater, solar or geothermal), distribution piping (includes pipe laid out in a grid system, manifold, and circulation pump), and controls.

Benefits of radiant heating:
- Uses little electricity
- Usually more efficient than baseboard and forced-air heating systems
- Minimizes the effect on residents with airborne allergies
- Minimizes the need for humidification of the air
- Heating is concealed and sound-free

Plastic piping for radiant heating

PEX is the preferred piping material for this technology. Compared to other plastic piping, it has proven to be durable and cost-effective. Typically the PEX piping used in this application has an integral diffusion barrier to prevent oxygen permeability. CPVC, composite, PE-RT, and PP piping can also be used.

Figures 28 and 29
PEX tubing layout system
for radiant heating.

Solar Water Heating

According to the U.S. Department of Energy, water heating each year is the second largest expense in a home—approximately 14 percent to 18 percent of utility bills. Currently, solar hot water heating is the most effective way

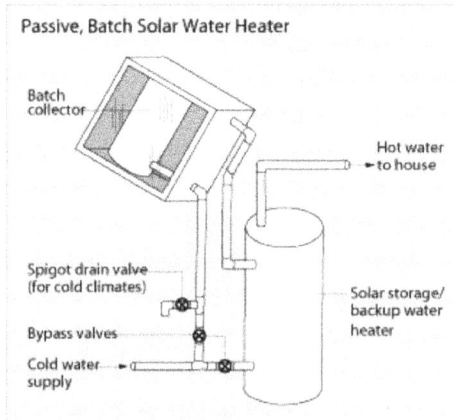

Source: U.S. Department of Energy. Energy Savers: Solar Water Heaters.

Figure 30 Passive batch solar hot-water heater.

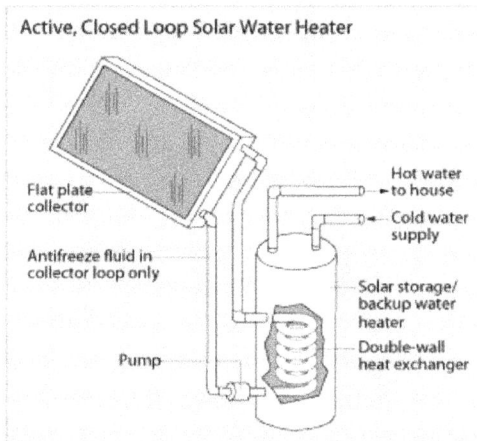

Source: U.S. Department of Energy. Energy Savers: Solar Water Heaters.

Figure 31 Active closed-loop solar hot-water heater.

to harness energy from the sun to heat water for the home.

Solar systems consist of collectors or solar panels to collect energy from the sun. The collected heat then warms the cold water as it flows into well-insulated storage tanks. The pre-heated water is then distributed to existing water heaters and/or radiant space heaters.

Solar water heating systems can be active (using circulation pumps and controls) or passive (without pumps or controls). Active systems are more energy efficient than passive; however, the passive system costs less and lasts longer. Choosing which system to use is based on the climatic area of the building site and the cost of local energy providers.

Figure 32 Roof-mounted solar panels.

Benefits of solar water heating

- Can last up to 50 years
- Low operating costs
- Minimal maintenance
- Reduces energy consumption
- No greenhouse gases or other harmful emissions

Plastic piping for solar water heating

Piping that is used in hot-water distribution systems are also candidates for use in solar water heating. This would include CPVC, PEX, PE-RT, PP, and composite piping.

Water Conservation

According to the University of Michigan, 97.5% of Earth's water is salt water, with the remaining 2.5% being fresh water. However, nearly 70% of Earth's fresh water is locked up in Antarctica and Greenland. The remaining fresh water exists as soil moisture or in inaccessible underground aquifers. Less than 1% of the world's fresh water is accessible and sustainable for direct human use. Today over 1 billion people do not have access to potable water. By 2025, it is estimated that over 3 billion people in 90 different countries will face severe water stress.

Although trenchless piping and desalination are not directly used in building construction, these technologies were added because they have a profound effect on the entire infrastructure of providing potable water to populations throughout the world.

The following technologies assist in protecting and conserving our precious freshwater:

- Gray Water Reuse
- Water Efficient Irrigation
- Rainwater Harvesting
- Trenchless Piping
- Desalination

Figure 33 Water reservoir.

Gray Water Reuse

The typical American family of four uses 400 gallons of water each day (this is about twice the usage amount of Europeans and seven times that of African families). Of that amount, 70% is used indoors and the rest mainly for outdoor irrigation. If a significant portion of indoor water could be reused—recycled essentially—it would greatly help to conserve water. Gray water is the wastewater from bathroom sinks, bathtubs, showers, and clothes washers. Gray water accounts for about 60% of home wastewater. It contains far fewer pathogens and nitrogen than black water (toilet flushing, kitchen sinks, and dishwashers).

Generally, gray water reuse systems collect and move gray water in a dedicated waste piping system, through a three-way diverter valve, and into a sand filter or settling tank that serves to remove course materials. The resultant fluid is retained in a non-pressurized storage tank, and then transferred under pressure to non-potable outlets for use in code-approved applications such as sub-surface irrigation, flushing toilets, and exterior washing.

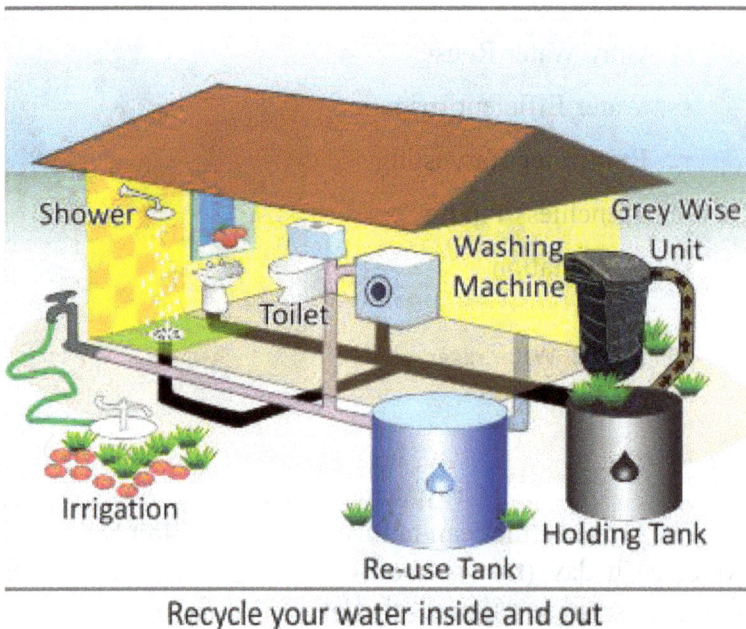

Recycle your water inside and out

Figure 34 Piping schematic of gray-water reuse.

Benefits of gray water reuse

- Water savings estimates are as high as 30% to 40% of yearly water bill

- Reduced costs for wastewater generated onsite

- Reduced costs for hauling or emptying sewage tanks

- Reduced overall energy-use costs for conveying and treating water

- Less capital investment of wastewater conveyance and treatment facilities

Plastic piping for gray water reuse

Purple-colored, pressure-rated PVC piping has the market share for this application. Purple is used as a signal to others that the piping system is not designated to handle potable water. For applications, handling gray water that might exceed 140° F/60°C, purple-colored CPVC and PEX piping is available.

**Figure 35
PVC gray-water
reuse piping
(non-potable water).**

Water-Efficient Irrigation

Water-efficient irrigation (sometimes referred to as drip, micro, low-flow, low-volume, or trickle irrigation) applies the correct amount of water required in a landscaped area, exactly when and where required, with minimal wastage.

This type of irrigation system delivers the water as near as possible to the roots of the plant, with the largest droplet size possible. A properly designed water-efficient irrigation system can reduce water usage up to 75% because it reduces water evaporation and runoff.

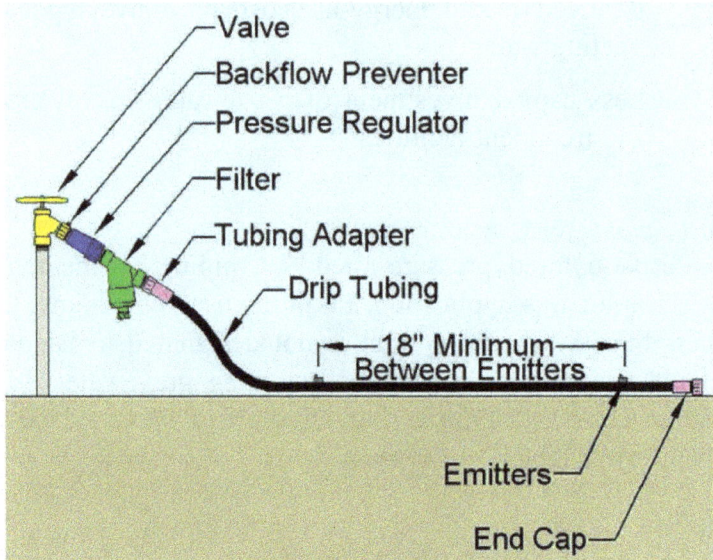

Source: Stryker.

Figures 36 Typical schematics for low-drip irrigation systems.

Typically, drip irrigation consists of a pressurized water source connected to a valve, backflow preventer, pressure regulator, filter, tubing adapter, drip tubing, and, last, emitters and end caps. In addition, devices such as pressure regulating sprinkler heads, controllers, automated shut-off valves, and rain sensors are used to reduce over watering.

Benefits of water-efficient irrigation
- Uses less water by regulating the exact amount of water to a zone or plant

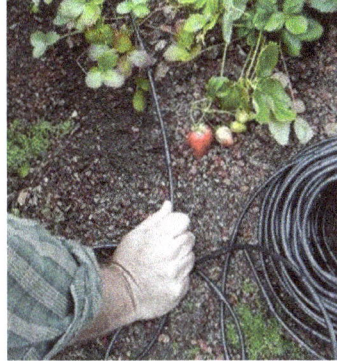

Figures 37 and 38 PE tubing for water-efficient irrigation systems.

- Reprogramming possible as the landscape matures and changes

- No loss of water due to wind or sun evaporation

- Minimizes foliage disease

- Water slowly soaks into the soil, maintaining better soil structure

Plastic piping for water-efficient irrigation

For low-drip irrigation, with few exceptions, polyethylene is the piping material of choice. In many cases, larger diameter PVC piping and pumping systems are used to bring water to the smaller diameter PE tubing network for final distribution.

Rainwater Harvesting

The collection or "harvesting" of rainwater has been in use for thousands of years. Depending on local codes, harvested water can be used in gray water applications such as irrigation, toilet flushing, and laundry washing or, in some locations, as bathing and drinking water (following appropriate treatment).

The components for rainwater harvesting are: catchment or collection surface areas such as roofs, gutters, or rainwater piping conveyance systems; holding tanks or cisterns; and delivery systems using gravity or pumps with piping and controls. Depending on the applications, treatment systems can include filter and reverse osmosis devices.

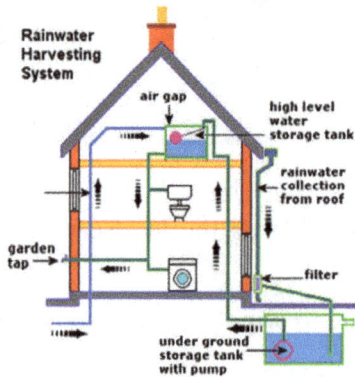

**Figure 39
Typical residential
rainwater harvesting
piping system.**

Benefits of rainwater harvesting
- Savings on annual water bill can be as high as 30% to 40%

- Free water once the initial investment is recovered

- Can construct buildings in areas not serviced by water sources

- Assists in recharging and improving groundwater supply

- Allows for plant growth and environmental stability

Plastic piping for rainwater harvesting
PVC is the most commonly used piping material in rainwater harvesting, although ABS could also be used.

Source: Rainwater Connection

Figure 40 PVC piping used for rainwater harvesting systems.

Trenchless Piping

According to the PVC Pipe Association, corrosion is the major cause of more than 850 water main breaks per day in North America. A staggering fact: these breaks result in a loss of 2.6 trillion gallons of potable water (17% of pumped drinking water) each year.

In the past 30 years, trenchless piping has emerged as the technology of choice to replace and/or rehabilitate failed underground piping systems with minimum disruption of above ground traffic, roads, or buildings. This technology can be executed by the use of three proven processes: slip-lining, pipe-bursting, and horizontal drilling (sometimes called directional drilling).

Slip-lining, normally installed in larger diameter piping systems, uses a flexible plastic pipe or liner that is inserted into and pulled through the walls of the failed piping system. In most cases, the smaller-diameter liner has much better flow characteristics than the original piping, and will not reduce the specified design flow of the original piping.

Figure 41 Plastic piping replacing corroded concrete sewer piping.

Figure 42 PE pipe slip-lined in failed municipal sewage piping system.

Pipe-bursting uses a pneumatic drilling apparatus whose drill bit follows and bursts the failed existing underground piping, while axially pulling along the same or a larger diameter flexible pipe. In most cases, smaller-diameter piping (below 12 inches) is ideal for this technology.

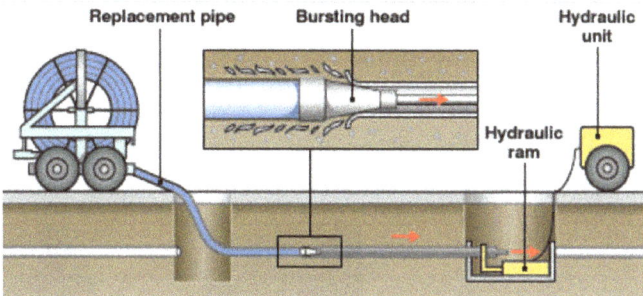

Figure 43 Pipe-bursting schematic.

Horizontal drilling starts with the drilling a vertical hole, and then the drill is shifted into a horizontal plane to continue drilling. Flexible plastic piping can then be pulled right behind the drill bit through the casing. This system enables piping to be installed underground, incorporating the benefits of the other trenchless piping methods.

Figure 44 Horizontal drilling rig

Benefits of trenchless piping
- More timely repairs of underground failed piping systems

- Less expensive than totally replacing failed piping

- Minimum, if any, disruption of aboveground structures or traffic

- Replacement piping will normally last longer and perform better

- Technologies are proven and have been successful for decades

Figure 45
PE piping being installed in horizontal drilling application

Plastic piping for trenchless piping
The dominant piping materials for the three forms of trenchless piping are PE and PVC.

Desalination

Presently, the United States has over 1,200 operating desalting plants. Desalting of water has been going on for centuries, mainly through the use of evaporative-distilling.

There are three types of distilling processes in use today: multistage flash evaporation (MSF), multieffect distillation (MED), and vapor compression (VC). Very large amounts of energy are needed for these processes; to operate more cost effectively, distilled desalination plants are commonly linked to cogeneration plants.

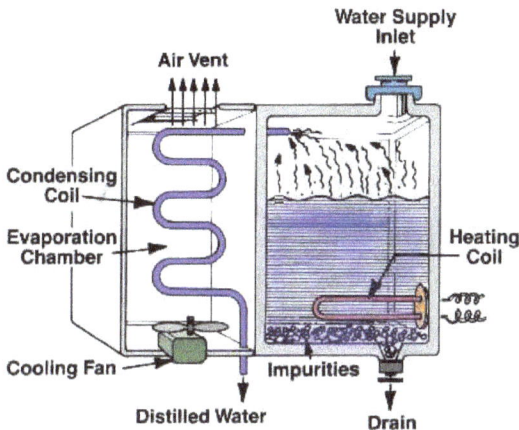

Figure 46 Distillation schematic for evaporative desalination.

The membrane process was introduced in the early 1960s. Today it is the primary method of producing freshwater, especially in very large desalination plants. There are two types of membrane processes: electrodialysis and reverse osmosis (RO). Reverse osmosis is becoming the preferred desalination processes, for its durability, modular design, and cost-effectiveness. This technology uses a specially-designed membrane. Saltwater on one side of the membrane is forced under high pressure through the plastic membrane, diffusing the freshwater into a down-line treatment process. The remaining concentrated salt solution is then disposed of in an environmentally-sound manner. Several RO desalination plants now being completed or on the drawing boards will be able to produce over 300 million liters per day of freshwater.

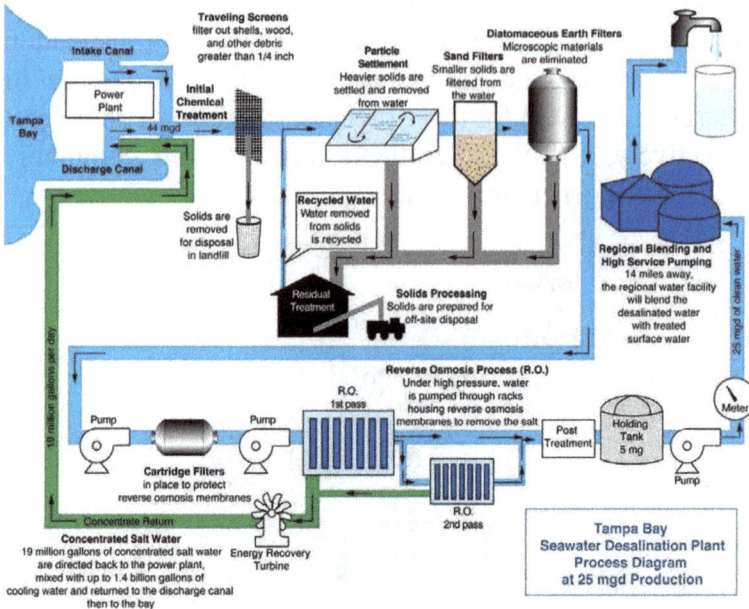

Figure 47 **Reverse osmosis desalination schematic.**

Benefits of desalination

- Provides potable water from one of earth's most abundant resources—seawater

- Desalination is sustainable

- Freshwater scarcity in parts of the world relies on desalination to fill the void

- As freshwater becomes more costly, desalinated water is more affordable

- May be the only feasible method of providing drinking water to future populations

Plastic piping for desalination

PVC and PE piping are used extensively in seawater intake and outlet lines. Low-pressure pre- and post-treat-

Figure 48 Large desalination plant using RO processing.

ment as well as freshwater storage and distribution commonly use PVC piping.

Summary

For your convenience and review, we have included in this reading's appendix a brief description of commonly used plastic piping materials, plus Table 1, which lists the most commonly used plastic piping systems for each green building technology.

As green technologies develop and are implemented, plastics will remain at the core of sustainable building solutions that improve the quality of life and health, save energy and conserve water. Plain and simple, plastic piping has a lot to recommend it:

- *Very durable*: minimum corrosion and is engineered for long service life

- *Saves energy*: uses less energy to manufacture, install, use, and recycle resulting in lower greenhouse emissions

- *Offer affordable sustainability*: has lower material, installation, and maintenance costs

- *Enables sustainable systems*: saves lives, water and energy

- *Contributes to green building certification*: in interior and exterior building system

Plastic piping is a crucial building block of green building construction's evolution. With this evolution, the plastic piping industry will continue its focus of achieving the ultimate goal: to save lives, energy, and water.

Notes:
1. Special thanks and recognition for much of the source materials used in this presentation goes to authors Jabeen Quadir and Abraham Murra of Sustainability Edge Solutions for their report, *Green Building Technologies that Use Plastic Pipe and Tubing to Function*; also to Plastic Pipe and Fittings Association (PPFA) and Plastic Piping Education Foundation (PPEF), which sponsored Quadir and Murra's report.
2) For more detailed information on plastic piping systems please refer to the following websites:
   - Plastic Pipe and Fittings Association: www.ppfahome.org
   - Chasis Consulting: www.sustainablepipingsystems.com
   - Plastic Piping Institute: www.plasticpipe.org
   - PVC Pipe Association: www.uni-bell.org
   - OPUS: http://opus.mcerf.org

Appendix
Types of Plastic Piping Materials Commonly Used in Green Building

ABS (Acrylonitrile Butadiene Styrene): Primarily for drainage use, ABS has a 20 to 30°F higher temperature range than PVC. To its advantage, it normally requires just a one-step method of solvent cementing. ABS also has excellent material characteristics in cold weather.

Composites (PE x AL x PE and PEX x AL x PEX): This material refers to tri-layered piping consisting of a central aluminum or steel extrusion with an inner and outer wall of either PE or PEX. Composites were introduced to the market decades ago; although non-composites successfully handle higher temperature water distribution at a lower installed cost, tensile and burst strength test results favor composites.

CPVC (Chlorinated Polyvinyl Chloride): Its advantage is that it has most of the excellent material characteristics of PVC, but can handle temperatures that are 70°F higher (210°F/99°C max). This feature makes CPVC effective in handling residential hot water applications and in meeting the strict residential fire sprinkler system codes.

PE (Polyethylene): This is a very flexible material and has the added benefit of being abrasion and chemically resistant. Most of its uses are in belowground applications, such as sewer lines, geothermal ground loop systems, trenchless piping, and low-drip irrigation. The breadth of pipe diameter sizes in PE is vast, beginning at micro-sizes and going up to five feet or more.

PE-RT (Polyethylene-Raised Temperature): One of the newest piping materials to enter the plumbing market, PE-RT has many of the features and benefits of PE, plus it can handle temperatures that are more than 50°F higher. PE-RT is now being used in hot water distribution piping. Either mechanical joining or heat fusion methods can be used to install this piping system.

PEX (Polyethylene cross-linked): One of the fastest growing plumbing piping materials in North America, PEX is gaining market share in hot-water pressure distribution installation. PEX success is based on the following features: fewer fittings required, color coded, available in coils exceeding 1,000 feet in length, and easy to join.

PP (Polypropylene): A broad range of PP polymers and copolymers plus specially designed materials using a fiberglass composite allow PP to be used in hot-and cold-water distribution, chemical waste DWV, process pipe, and high purity water applications.

PVC (Polyvinyl Chloride): Commonly used for piping applications where conditions of service are below 140°F/60°C and less than 150-psi of working pressure, PVC is the dominant player. Estimates are that PVC makes up over 70% of all plastic piping systems in use today. PVC is the most widely-used plastic in building construction. The reasons: good tensile strength; breadth of product line; easy to install; chemically resistant to most acids, bases, and salts; cost-effective.

Table 1 Green Technologies and Plastic Piping

Direct Impacts	Plastic Pipe Materials[1]
Life and Health	
Central vacuum systems	PVC/ABS
Double containment systems	All
Foundation drainage	PVC/ABS
Radon venting	PVC/ABS
Residential fire sprinkler	CPVC/PEX
Energy Savings	
Decentralized wastewater treatment	PVC/ABS/PE
Drain-waste heat recovery	CPVC/PEX/PE-RT/PP/Composite
Geothermal energy	PE/PEX/PP/Composite
High efficiency hot water distribution	CPVC/PEX/PE-RT/PP/Composite
Radiant floor heating	PEX/PE-RT/PP/Composite
Solar water heating	PE/PEX/CPVC/PE-RT/PP/Composite
Water Conservation	
Efficient irrigation	PE/PVC
Gray water reuse	CPVC/PVC/PEX
Rainwater harvesting	PVC/ABS/PE
Trenchless piping	PE/PVC
Desalination	PE/PVC/PEX/PP/Composite

Note 1: Piping material abbreviations:
　　　ABS—Acrylonitrile Butadiene Styrene
　　　Composite—PE x Aluminum / steel x PE & PEX x Aluminum / steel x PEX
　　　CPVC—Chlorinated Polyvinyl Chloride
　　　PE—Polyethylene
　　　PE-RT—Polyethylene - raised temperature
　　　PEX—Polyethylene Cross-Linked
　　　PP—Polypropylene
　　　PVC—Polyvinyl Chloride

THINK PLASTICS—THERMOPLASTIC PIPING
GAINING INDUSTRY ACCEPTANCE

When English poet Thomas Gray wrote in 1742 that "ignorance is bliss," he could not have foreseen how U.S. industry would stand that phrase on its head during most of the last half of the 20th century.

A lack of knowledge by many design and plant engineers has kept many U.S. manufacturers in the dark during the past forty years as to the many tangible benefits of thermoplastic piping systems over less efficient and more costly metal products. Ignorance, in the case of manufacturing, has resulted in lost time and wasted money.

While plastic piping products have become the dominant material in many markets in recent decades—including plumbing, drainage, gas transmission, electrical conduit, acid waste drainage, water lines, and underground irrigation—the bulk of manufacturers have been slow to embrace the use of thermoplastic piping systems.

The advantages of plastic piping compared to other materials are well documented. Also, the overall cost of thermoplastic products is below that of metal valves, fittings, and pipe. Estimates are that plastic pipe, fittings, and valves are capable of handling up to 70%, of all industrial applications. Yet, thermoplastics account for less than 15% of dollar volume purchases in the industrial market.

Figure 1
Preparing to join 24-inch
diameter PVC pipe.

Why has U.S. industry been slow to join the plastics revolution? Several reasons come to mind. First, those of us in the thermoplastic piping products industry must take the blame for not having done a very good job of educating the marketplace or the public about the benefits and capabilities of thermoplastic products.

Second, a minimum of research and development in industrial plastic piping products has been performed. As a result, few new products or piping materials have been developed domestically; most are coming from importers.

Third, to a large degree, U.S. industries have been unwilling to change their habits and adapt to more progressive piping materials. Industries such as pulp and paper, utilities, oil and gas, petroleum and ore refineries, and steel manufacturing could greatly benefit from increased use of thermoplastic piping systems. A lack of economic pressure in past years also has kept these industries from moving toward plastic materials, but today's global competition will likely bring about some change.

What can be done to increase the industry's use of thermoplastics? A few suggestions and observations:

• Plastic piping product and resin manufacturers should take a united approach to educate users, engineers, contractors, and students on the merits of designing plastic piping systems.

- American innovation and research into plastic industrial piping products must he stepped up.

- American industry may eventually "wake up and smell the coffee," as newspaper columnist Ann Landers would say. Use of thermoplastics, the most effective and cost-saving piping material, can help to ensure the worldwide competitiveness of U.S. industry.

Advantages of Thermoplastics

Because the advantages of plastic piping are significant, the tide of opinion within the manufacturing industry should eventually turn in favor of its use. Thermoplastic features can produce considerable cost savings while increasing piping system reliability. Advantages include:

- **Corrosion resistance.** Plastics are nonconductive and are immune to galvanic or electrolytic corrosion, which is a major cause of metal pipe failure. Because the outer wall of plastic pipe is corrosion resistant, plastic pipe can be buried in acidic, alkaline, wet, or dry soils with no paint or special protective coating applied.

Figure 2 Corroded metal valve.

- **Chemical resistance.** The variety of materials available allows almost any chemical, at moderate temperatures, to be handled successfully by plastic piping.

- **Low thermal conductivity.** All plastic piping has low thermal conductance. This feature maintains more uniform temperatures in transporting fluids in plastic piping than in metal piping. Minimal heat loss through the pipe wall of plastic piping may eliminate or reduce greatly the need for pipe insulation.

- **Flexibility.** Thermoplastic piping materials are relatively flexible. Piping flexibility is a major asset, particularly in underground piping installations. Plastic piping may be "snaked" in trenches to minimize the effects of expansion and contraction; underground pipe bends may be used more readily with plastics eliminating or minimizing the use of pipefitting ells and ensuring a more dependable piping system.

- **Low friction loss.** The interior walls of all plastic piping have a Hazen and Williams C factor of 150 or higher, generally resulting in less horsepower required to transmit fluids in plastic piping compared with metal and other nonmetallic piping. This feature allows small diameter piping to be used in place of larger non-plastic piping systems, resulting in cost savings to the user.

- **Long life.** Very little change occurs to the physical or molecular characteristics of plastic piping over dozens of years of use. Theoretically, in most installations of plastic piping, there is no known end-life of the piping system.

**Figure 3
Smooth inner walls of
plastic piping.**

- **Lightweight.** Most plastic piping systems are a minimum of one-sixth the weight of steel piping. This feature results in less freight and installation costs. Lightweight plastic piping systems are easier to install in close quarters and, in many cases, underground piping requires no expensive heavy lifting equipment.

- **Variety of joining methods.** Thermoplastics can be cemented, heat-fused, threaded, flanged, and compression fitted. A new push-to-connect fitting on the market today makes the assembly of plastic piping systems for quicker and easier by eliminating the need for conventional threaded or solvent-cemented connections. The variety of joining methods allows plastic piping to be adapted easily to any field application.

- **Nontoxic**. Plastic piping systems are nontoxic and odorless. Additionally, thermoplastics are resistant to biological attack, abrasion, and weather. When all of the above features of plastic piping are considered, substantially less cost is required for use of thermoplastic piping systems.

Design Considerations

Distinct differences exist in designing with and using thermoplastics. Depending on the application, engineering specifiers should be aware of the fact that plastic piping systems have much less heat resistance than metals. When pressure-piping applications exceed temperatures of 300°F/149°C, the use of solid plastic piping systems is very limited. Engineers must determine the installation's maximum external and internal temperatures when selecting the piping material.

In some cases, not designing for the piping external temperature could cause excessive sagging because of a lack of pipe supports. When designing with thermoplastics, the nature of the material must be taken into account: When the temperature increases, the tensile strength and working pressure decrease. It is also important to keep in mind that expansion and contraction of thermoplastics is greater than with metals. Anchors and expansion devices must be considered in the system design for proper performance.

Other considerations when using thermoplastics are the need for more pipe supports, and the fact that plastics are less impact resistant than metal systems. Contractors also may lack expertise in joining thermoplastics, which could result in a loss of system integrity.

The key to success with thermoplastic piping systems is product knowledge. Manufacturers and distributors of thermoplastic piping systems must help to increase awareness and understanding of plastics within industry.

Billiard Balls

The actual history of plastics can be traced back to the scarcity of ivory in the manufacture of billiard balls. In 1869, inventor John Wesley Hyatt of Starkey, NY, developed plastic celluloid as an ivory substitute by mixing pyroxylin, camphor, and nitric acid. Hyatt had a long career as an inventor, developing a water filter, a multiple-needle sewing machine, and the Hyatt roller bearing, but celluloid is his best-known discovery[1] and is regarded as the first important plastic.[2]

Plastic piping applications date from the mid-1930s when PVC piping was first used, as it still is, for sanitary drainage in Germany. Following World War II, polyvinyl chloride, polyethylene, and reinforced plastics were introduced into many industries. Germany and Japan embraced use of plastics for piping, especially thermoplastics, which required a minimum of investment capital.

Use of plastics in piping started slowly in the United States. Wide public acceptance of plastics was not obtained until the late 1950s and early 1960s. Since that time, the use of thermoplastic and thermosetting piping products has increased at an astounding rate. Plastic piping materials are now used in almost every industry for all conceivable applications.

The most commonly used plastic piping applications are non-industrial, such as irrigation piping, drain-waste-vent piping for homes, water mains and service lines, natural gas transmission service lines, and home fire and lawn sprinkler systems. In these markets, plastic piping is displacing metals as the preferred material.

Applications

While there is no limit to the possible new applications for plastics, the following examples serve to demonstrate the varied use of thermoplastic piping systems in industry.

**Figure 4
PE process piping.**

- **Food processing.** Most plastic piping materials are approved by the National Sanitary Foundation and receive Food and Drug Administration approval when required. The purity of the end product in any food-processing application is critical, and plastics fit the bill beautifully.

- **Plating.** The automotive, aircraft, electrotyping, and canning industries use thermoplastic piping where possible in their plating processes. Plastics are a natural in this market because almost every metal-salt plating solution can be handled easily, including brass, cadmium, chrome, copper, gold, lead, nickel, rhodium, silver, tin, and zinc.

- **Steel mills.** Ironically, steel mills are replacing steel piping with plastics. The mills realized that their manufacturing costs improved with use of plastics because of reduced maintenance, lower material costs, and longer life provided by plastic piping products.

- **Pulp and paper.** These plants handle four types of media: liquids, steam, water, and stock. Except for steam, plastic piping handles most of the other fluids under 275°F/135°C and 150 psi.

- **Electronics.** The manufacturers of solid-state electronics products such as semiconductors, rectifiers, and printed circuitry demand ultra-pure water to clean their products and prevent contamination. Thermoplastics are the preferred materials for handling ultra-pure water in which an ion exchange or demineralization system is employed. PVC, CPVC, polypropylene, PVDF, and other fluorocarbons are used for ultra-pure water distribution systems. Thermoplastics also are used for handling etching media such as sulfuric, nitric, hydrochloric, and hydrofluoric acids. Wastewater and air-handling systems required in the electronic industry also use plastic piping throughout.

- **Photographic laboratories.** All manufacturers of photographic process equipment and photographic chemicals specify and use thermoplastics.

- **Mining/oil/gas.** Plastic pipe, itself a derivative of oil and natural gas, have successfully been applied in handling all crudes, salt water, and natural gases. Most commercial gas companies today use millions of feet of plastics in natural gas distribution. Polyethylene piping, colored beige, yellow, or orange, is the preferred material for this application. In the mining industry, the most popular use of thermoplastics is in ore leaching, in which the ore is treated with dilute sulfuric acid or sulfides and then with ferric sulfate solutions. PVC, CPVC, ABS, and polyethylene piping are used in many of the leaching process stages. Plastics also are used for the movement of slurries and the handling of acids.

- **Marine applications.** Shipbuilding, marinas, fish hatcheries, marine research, and amusement/theme parks are using significant amounts of plastic piping. The Aquarium of the Pacific in Long Beach, CA, and the Wet-n Wild water park in Rio de Janeiro, Brazil, are two recent examples in which thermoplastic piping systems were installed. From advanced atomic submarines to shrimp boats, thermoplastic piping is used on board ships for water and brine applications.

- **Sewage treatment.** Whether in primary or secondary treatment phases, plastics are used throughout water and sewage treatment facilities.

Figure 5 PVC X stainless steel double containment manifold.

Additional applications for plastic piping systems include power plants, plumbing, heating/air conditioning/refrigeration, institutional facilities, and heavy construction. In fact, uses for plastic piping systems seem to be limited only by one's imagination.

"Plastics," the famous one-word line from the 1967 film *The Graduate*, was said to be the key to the future. That prediction was never truer than it is today for U.S. industry.

References
[1]Biography.com, A&E Television Networks.
[2]The Concise Columbia Electronic Encyclopedia, 3rd Ed.

Portions of this article were adapted from the book *Plastic Piping Systems*, 2nd Ed., by David A. Chasis, Industrial Press.

32

TRENCHLESS PIPING AND PLASTICS [1]

According to the PVC Pipe Association, corrosion is the major cause of more than 850 water main breaks per day in North America. A staggering fact: these breaks result in a loss of 2.6 trillion gallons of potable water (17% of pumped drinking water) each year. Polyvinyl Chloride (PVC), a critical part of the solution, is getting due recognition with the endorsement by *Engineering News Record* of PVC piping for water and sewer, and its further acknowledgment of PVC as one of the top 20 engineering advancements in more than a century. Further confirmation by the American Water Works Association Research Foundation found the life expectancy of PVC water main pipe to be in excess of *110 years*.

PVC and Polyethylene (PE)—PE also exhibits high performance and long service life—have become the materials of choice for new and replacement municipal water mains and sewer installations. [Other plastic piping materials to a lesser degree may also be used in this market such as Polypropylene (PP), Acrylonitrile Butadiene Styrene (ABS), Polyethylene Cross-linked (PEX), and Chlorinated Polyvinyl Chloride (CPVC).]

In the past 30 years, trenchless piping has emerged as the technology of choice to replace and/or rehabilitate failed underground piping systems with minimum disruption of above ground traffic, roads, or buildings. This technology

Note [1]: Much of the copy and photos were excerpted from a David A. Chasis article to be published, *Sustainable Piping Systems for Green Building*.

can be executed by the use of three proven processes: slip-lining, pipe-bursting, and horizontal drilling (sometimes called directional drilling).

Slip-Lining

Slip-lining, normally installed in larger diameter piping systems, uses a flexible plastic pipe or liner that is inserted into and pulled through the walls of the failed piping system. In most cases, the smaller-diameter liner has much better flow characteristics than the original piping, and will not reduce the specified design flow of the original piping.

**Figure 1
Plastic piping replacing corroded concrete sewer piping.**

Pipe-Bursting

Pipe-bursting uses a pneumatic drilling apparatus whose drill bit follows and bursts the failed existing underground piping, while axially pulling along the same or a larger diameter flexible pipe. In most cases, smaller-diameter piping (below 12 inches) is ideal for this technology.

**Figure 2
PE Pipe bursting through vitrified clay piping.**

Horizonal Drilling

Horizontal drilling starts with drilling a vertical hole; next, the drill is shifted into a horizontal plane to continue drilling. Flexible plastic piping can then be pulled right behind the drill bit through casing materials. This system enables piping to be installed underground, incorporating the benefits of the other trenchless piping methods.

Figure 3
Horizontal
drilling rig.

The benefits of trenchless piping versus other piping systems are obvious and include:

- More timely repairs of underground failed piping systems

- Less expensive than totally replacing failed piping or open trenches for new piping

- Minimum, if any, disruption of above ground structures or traffic

- Replacement piping will normally last longer and perform better

- Technologies are proven and have been successful for decades

**Figure 4 PE piping slip-lined in failed municipal
sewage piping system.**

According to Bonner Cohen in his April 2012 publication, *Fixing America's Crumbling Underground Water
Infrastructure,* America will need to spend $3 to 5 *trillion*
dollars in the next twenty years to upgrade water and waste
treatment systems. Just to build and replace water and sewer
lines alone during this period will cost $660 billion to $1.2
trillion. As more and more water and waste-treatment utilities explore piping materials to meet future needs, plastic
piping without a doubt will be one of the leading materials
of choice.

Reprinted with permission of the IAPD; issue december 2012/january
2013 – **the IAPD magazine**

WHY PLASTIC PIPING?

Two major criteria for purchasing a car—or for that matter any non-commodity product—are performance and cost. This should be the exact same parameters when choosing piping systems: performance and cost. Yet in the piping mythology, adhered to by some unscientific and myopic agenda activists, these two major factors are laid aside or treated as insignificant. Maybe the reason for this is the need today, or at least it seems, to bash or find fault with any hugely successful company or product. Certainly this could be a major reason for all the negativity in recognizing the unbelievable revolutionary success of thermoplastic piping during the last fifty years.

Through out the world, Polyvinyl Chloride (PVC), Polyethylene (PE), Polypropylene (PP), Chlorinated

Figure 1 Plastic pipe lightweight makes installation easier and safer

Polyvinyl Chloride (CPVC), Cross-linked Polyethylene (PEX), and Acrylonitrile-Butadiene-Styrene (ABS) piping systems have made tremendous inroads in such applications as residential and commercial drain-waste-vent, hot- and cold-water distribution, chemical/acid drainage systems, irrigation systems, swimming pools, well casing, natural gas distribution, and several others. Why, in just five decades, have plastics overtaken metal, asbestos-cement, and clay pipe in many applications? Right, you guessed it—performance and cost.

Figure 2 Plastic piping can be made in any color imaginable.

Let me explain. In most cases, plastics cost less than other piping materials, but the real cost savings is that plastic piping is easier to install (lightweight), maintain (almost no corrosion or chemical attack), and has less onsite thefts. Plastics are a non-conductor and require no or minimum insulation. All thermoplastic piping systems have optimum flow characteristics and are more abrasive resistant than most other piping systems. These characteristics allow plastics to have less friction loss in the pipe, allowing the need for less energy to convey fluid from point A to B; due to the abrasion resistance characteristic, plastics in most cases can handle fluids at higher velocities than non-plastic piping systems. This attribute can reduce the pipe diameter size, saving piping costs and, in the case of hot water distribution, realize a savings in water usage.

Plus, the joining systems for plastic piping produce a homogenous, monolithic joint—which means no intrusion or extrusion of fluids. A recent study of municipal piping pointed out, for example that based on equal amounts of cast iron, ductile iron, asbestos-cement, and plastic piping, plastics showed it was head and shoulders over other pipe materials and accounted for just a fraction over *1% of all repairs*. On top of all these benefits, thermoplastic piping has excellent sustainability regarding issues of economy, energy, and environment.

Figure 3 Plastic pipe can be easily fabricated, as shown in PP manifold.

Figure 4
Plastic pipe can be coiled,
as shown with PE natural
gas distribution piping.

So the bottom line is this: thermoplastic piping is environmentally sound, easy and safe to install, reliable, long-lasting, and cost-effective. These are just some of the reasons for the resounding success of plastics. Critics of plastics can harp on unproven reasons not to use these materials, but they ignore the main point—the market place has embraced plastics as one of its preferred piping materials.

Reprinted with permission of the IAPD; issue june/july 2008 – **the IAPD magazine**

PART 3
Plastic Materials and Products

PLASTIC PIPING MATERIALS AND PRODUCTS

These articles inform readers about the most common plastic piping materials available for use in dozens of applications. In addition, plastic valves, fabrications and flanges are also discussed:

34. *ABS...The Other Piping Material* explains in detail the wide spectrum of Acrylonitrile Butadiene Styrene (ABS) usage in many residential and commercial applications.

35. *CPVC Piping Systems* lists several applications in which Chlorinated Polyvinyl Chloride (CPVC) piping systems can handle fluid temperatures 70° to 80°F higher than Polyvinyl Chloride (PVC).

36. *Environmentally Sound Piping Systems* informs readers of piping systems designed to protect the environment and health of workers when dealing with harmful fluids.

37. *Facts Regarding PVC Piping and the Environment* highlights facts that refute many of the anti-PVC lobby's concerns with PVC piping systems.

38. *Homeowners Embrace PEX for Hot and Cold-Water Distribution* offers many reasons why Polythene Cross-linked (PEX) piping material is making tremendous inroads in residential home construction.

39. *Plastic Piping System Fabrications* demonstrates that if engineering designers can draw on paper their needed product, experienced plastic fabricators can construct it.

215

40. *PVC...Pipe's Most Versatile Material* lists dozens of Polyvinyl Chloride (PVC) products and applications, and explains why PVC piping, by footage, is one of the leading piping materials in the world.

41. *Recycling PVC Piping Systems* addresses the industry's efforts to maximize recycling of all PVC products before and after manufacturing of the end product.

42. *Super Pipe* describes a new generation of PVC pipe manufacturing methodology that improves the properties of standard PVC piping, offering new opportunities in the municipal water and other markets.

43. *Thermoplastic Flanges* details the hows and whys of plastic flange usage in commercial and industrial applications.

44. *The Ultimate Piping Material* gives examples why Polyvinylidene Fluoride (PVDF) piping systems are capable of handling the most challenging applications of almost any plastic or non-plastic piping material.

45. *Why PVC Piping Systems?* presents a solid argument that, if conditions of service allows, PVC is most likely the best piping material of choice.

46. *Why Plastic Valves?* provides a solid case that more plastic valves should be specified and installed, especially on plastic piping systems.

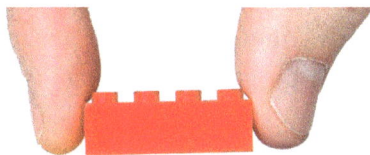

ABS: THE OTHER PIPING MATERIAL

ABS (acrylonitrile butadiene styrene) is the material of choice when it comes to the injection molding of many light and rigid products such as electronic components, automotive body parts, enclosures, protective head gear (football and motorcycle helmets), power tools, and toys. Ever hear of Legos?

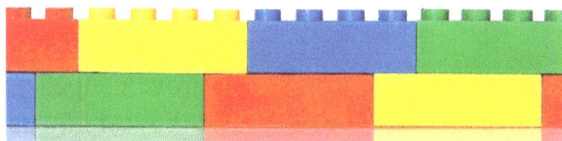

Figure 1 Legos made from ABS.

It's precisely the features of durability, ease of use, cost effectiveness, and environmental soundness that are needed for these product applications that make ABS an excellent piping material for a wide spectrum of residential, commercial, and industrial applications.

ABS was first discovered in the 1940s. By the 1950s it was being made into pipe. Although ABS might be at the top of piping's alphabet list, it has not yet gained the proper recognition or confidence of specifying engineers and builders in North America that it merits.

Durability

With over a half-century of successful usage and billions of feet of piping installed, ABS offers an impressive list of advantages similar to other plastic piping materials: excellent chemical, corrosion, and abrasion resistance; superior joint integrity; low thermal conductivity; and optimum flow characteristics. There's more—ABS also offers one of the broadest temperature ranges of any polymeric piping materials (–40°F/–40°C to 180°F/82°C). Even more, ABS has extremely high impact resistance throughout the complete temperature range, making it the DWV material of choice in many colder weather areas. And yet more—the presence of butadiene in the ABS polymer imparts greater ductility properties than many other piping products, which makes it an ideal material for burial in expansive soils (soils that migrate due to the presence or absence of moisture) and shifts in the earth due to seismic activity.

Easy and Safe to Install

There is nothing lighter in weight than ABS in the category of drain-waste-vent (DWV) piping systems. This light-weight feature is due to a proven pipe extrusion process known as *cellular core.* Over the last 25 years, several pipe manufacturers have developed a tri-layered ABS pipe designed for non-pressure applications. The inner and outer extruded layers are solid, rigid ABS. The middle sandwiched layer is composed of the same ABS resin, but injected with a blowing agent to create a larger cell structure, thus lowering the density of the center pipe layer. This design results in a lighter piping system that still maintains superior physical properties. Other advantages of ABS are the ease of installing one-step joining (no primer required), low VOC (volatile organic compounds) solvent cemented joints. It is one of the quickest to install and most leak-proof piping systems in the industry. ABS piping systems, unlike metal, do not use potentially dangerous open flames or molten lead for making joints; they eliminate the need for underground thrust blocking or seismic restraints; require no added coatings, wraps, or linings; and minimize the number of pipe joints with standard 10- and 20-foot long pipe lengths.

Figure 2 Cellular core section of ABS pipe

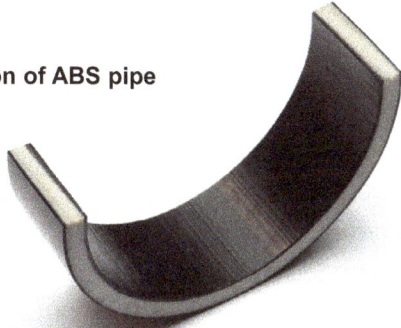

Cost Effective

A comprehensive study compared ABS to metal piping for DWV and storm drainage systems on a 12-story residential high rise. ABS piping systems showed a savings of over 80% in material costs and 25% in labor savings[1]. Additionally, the physical properties of ABS yielded further savings: huge savings in transportation costs (light weight); savings from eliminating external and internal pipe protec-

**Figure 3
Many more feet of
ABS pipe per truck-
load versus
metallic piping.**

tion and maintenance costs; savings from minimizing or eliminating expensive joining equipment and heavy-duty moving equipment (easy and safe to install); savings from having no field site thefts (low scrap value of ABS compared to most metals); and reduced insurance costs (minimizing field accidents due to ABS's light weight and lack of open flame or molten lead pots required for joining).

Environmentally Sound

Completed and preliminary Life Cycle Assessments (scientific studies usually prepared by a third party to determine the economic and environmental impact of manufactured products from cradle to grave) have shown that ABS piping systems compared to metals are the equal or better. From an energy use standpoint, ABS piping really shines compared to metals. It uses less energy in manufacturing; less in the transportation of finished products; less in the retention of fluid temperatures; less in the transportation of fluids (due to the smooth and long-lasting inner pipe bore); and less in recycling. ABS manufacturing processes of extrusion and injection molding eliminates almost all post-industrial scrap. And, when the product has completed its functionality, it can be easily and economically recycled into dozens of other useful products with many of the features and benefits of the original ABS material.

Applications

In North America, ABS is available in pipe diameters of 1¼- to 6-inch and is one of the preferred DWV piping materials in certain geographical regions, recreational vehicles, and manufactured housing. Other non-pressure ABS applications are: commercial DWV; drains for storm, roof, road and bridging; sub-service drainage; rain harvesting; decentralized wastewater treatment systems; gray water reuse, and radon venting. In Canada, Great Britain, and other European countries, ABS is used in pressure applications from ½- to 12-inch pipe diameters. These applications include: chemical processing, geothermal energy, HVAC, marine, mining, refrigeration, surface-finishing, swimming pools, and waste and water treatment systems.

Figure 4 ABS drain-waste-vent pipe residential installation.

Figure 5 ABS pressure pipe installation.

So you engineers and installers, consider how ABS can add all-around value to your future building projects. It has a lot to recommend it—durability, safety and ease of installation, environmental soundness, and cost-effectiveness. ABS:..the other piping material.

Note[1] : JB Engineering and Code Consulting, P.C.
Julius A. Ballanco, President
Report Number: 06A0706E1
July 6, 2006

Reprinted with permission of IAPD; issue april/may 2011 – **the IAPD Magazine**

35

CPVC PIPING SYSTEMS

Chlorinated Polyvinyl Chloride (CPVC) piping was introduced over five decades ago. It has been such a successful product that its installations now number in the hundreds of thousands. When first introduced, this new material was known by two terms: PVCC and High Temp PVDC (Polyvinyl Dichloride, which is chemically incorrect since it is not a dichloride). In Italy, CPVC is still referred to as PVCC and melodically pronounced "PV Chi Chi."

So what is CPVC? Basically, it is a resin. It's created by using a reactor to post-chlorinate PVC under prescribed pressure and temperature conditions. The purpose is to replace some of the hydrogen with chlorine in the vinyl chain. The resultant compound has physical material characteristics very similar to PVC, but with one important difference: CPVC can handle fluid temperatures that are 70°F, higher than PVC (PVC's maximum temperature limit is 140°F/60°C).

This new chlorine-enriched polymer opened up applications that PVC previously couldn't handle: hot-and cold-water distribution systems, reclaimed hot-water systems, fire suppression systems, and the transportation of process chemical solutions and chemical drainage at temperatures up to 210°F/99°C. The outstanding features and benefits of

PVC also apply to CPVC:

- Lightweight

- Does not support combustion

- Corrosion-, chemical-, impact- and abrasion-resistant

- Excellent flow characteristics

- Low thermal conductivity

- Variety of joining methods

- Code acceptance

- Integrated piping systems

- Ease of fabrication

- Environmentally sound

- Cost effective

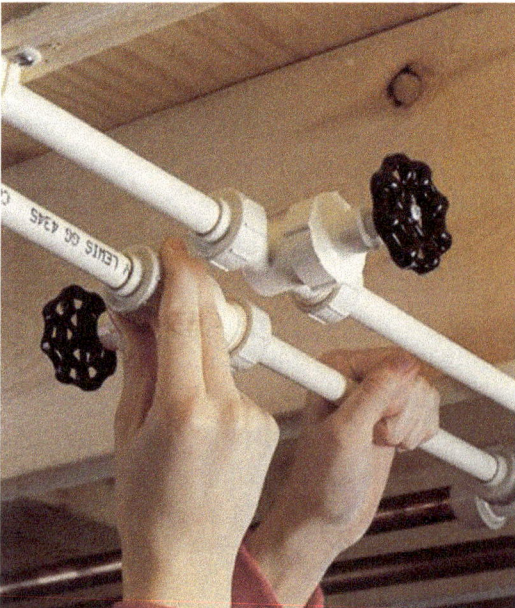

Figure 1 CPVC cooper tube-sized piping for hot-/cold-water distribution.

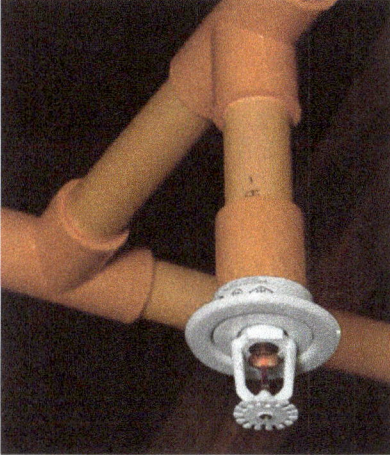

**Figure 2 CPVC fire sprin-
kler system.**

Another important feature distinguishing CPVC is the breadth of its product offerings. Pipe is available in CTS (copper tube size) up to 2-inch diameter and Schedule 40 and 80 from 1/4 to 24-inch diameter in lengths of 10 or 20-feet. Extruded duct is inventoried in diameters of 6 to 24-inches. Fittings, valves, pumps, sheet, rod, fans, scrubbers, and tanks are also available, offering a complete fluid handling system of just one material—CPVC.

**Figure 3 CPVC
scrubber packing.**

With this impressive breadth of line, there are few other piping materials that can handle residential and commercial potable hot-and cold-water distribution applications (including larger diameter vertical risers) nearly as well as CPVC. The same wide range of pipe diameters also offers a very broad line of chemical drainage systems, as well as ducting

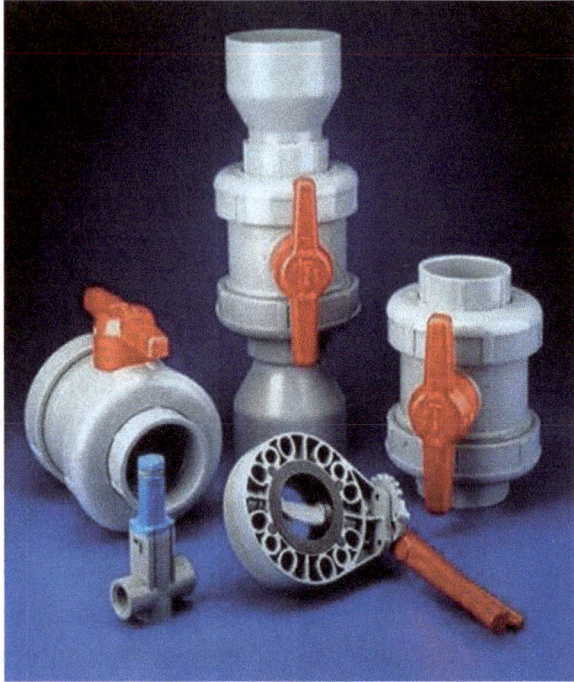

Figure 4 CPVC valves.

systems for corrosive environments.

How are CPVC piping products joined? Most installations use solvent cement joining techniques. Solvent cementing has been the preferred method of joining most vinyl piping products for decades. Why solvent cementing? Because these type of joints are easy and safe to make, are very durable, require no expensive tools, eliminate the need for open flames or heating plates, and are very cost-effective. Following the manufacturer's instructions, CPVC installers can solvent-cement joints in either a one- or two-step process. The one-step process consists of applying only the solvent cement to the properly prepared joining surfaces before joining. In the two-step cementing process, a primer is first applied to the joining surfaces and then followed by the solvent cement. Normally, thinner wall and smaller diameter CPVC piping require the one-step process. Always check with the manufacturer before choosing a particular cementing method.

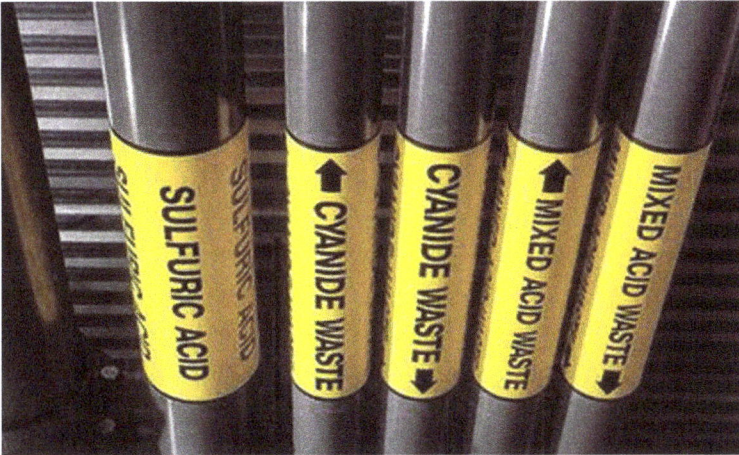

Figure 5 CPVC process piping.

CPVC is classified as a fire-safe material. This classification is based on CPVC's relatively high Limiting Oxygen Index (LOI) of 60, a Flash Ignition Temperature (FIT) of 900°F/482°C, and an Underwriter Laboratory (UL) flammability rating of 94V-0. Any LOI value of 21 or less (the approximate percentage of oxygen in the earth's atmosphere) will support combustion. If a flame is applied to CPVC, it will burn, but CPVC's relatively high LOI value ensures that once the flame is removed, the material will not support combustion; it will cease burning. CPVC's high FIT value is almost twice that of wood products. In other words, for CPVC to self ignite with a small external flame in the presence of a sufficient supply of combustion gas, the temperature has to reach approximately 900°F/482°C or more. The UL's flammability test measures a material's resistance to burning, dripping, glow emission, and burn-through. The 94V-0 designation is the highest category of resistance to burning of tested polymeric materials. Therefore, CPVC, in specific applications and with applicable building codes, could be used in return air plenums.

With its unique mechanical properties, especially higher temperature capability, CPVC is entering and capturing new markets outside of piping systems. The markets of residential siding, decking, window profiles, and fencing have all see increases in the use of CPVC. Also, combining CPVC as a substrate with the appropriate capstock material is allowing exterior building product manufacturers a wider array of offerings to their customers.

Therefore, when designing hot- and cold-water distribution; fire sprinkler, chemical drainage, and process piping systems; and exterior building products, consider the many advantages of CPVC before making a final decision.

Reprinted with permission of the IAPD; issue june/july 2012 – **the IAPD Magazine**

36

ENVIRONMENTALLY SOUND PIPING SYSTEMS

Two infamous names—Exxon Valdez and Deepwater Horizon—bring to mind the catastrophic damage man has inflicted on the environment. From the Exxon Valdez tanker spill, the oil industry learned the hard way that reconstructing tankers to a double-hull design was the best solution for preventing another devastating accident. Today's oil tankers have an inner carrier hull for carrying the oil as well as an outer hull to prevent and contain possible hull damage and leakage.

When an application calls for a virtual risk-free, no-leak piping system, the clear answer is double containment piping (DCP), known also as *dual containment*. This principle of dual containment has now been used for over three decades.

Most typical DCP systems have three basic components: carrier pipe, containment pipe, and spacers or centralizers (attachments normally placed every 4 to 5 feet and used to keep the containment pipe concentrically supported to the

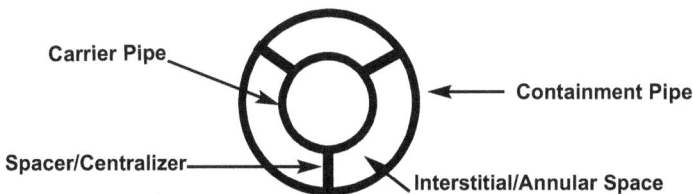

Figure 1 Nomenclature of typical DC piping.

229

carrier pipe). The gap between the inner and outer pipes is called the interstitial or annular space. To locate, pinpoint, and assess a carrier pipe leak, automatic leak detection sensors or visual sight glasses are employed.

In any DCP system, there must be room enough inside the containment piping to maneuver the carrier pipe and fittings. Table 1 lists the most common inner and outer pipe diameters. Let's look in closer detail at the three basic types of DCP systems: pre-assembled, custom-fabricated, and retrofit.

Table 1 Common Size Range of Double Containment Piping Systems

Carrier pipe Diameter (in.)	Container pipe Diameter (in.)
½	2
¾	3
1	3
1-1/2	4
2	4
3	6
4	8
6	10
8	12
10	14
12	16

Pre-Assembled DCP System

These systems are manufactured with stock lengths (usually 20 feet) of standard piping materials and pre-assembled fittings. Their advantages lie in the speed of shipping and lower overall cost.

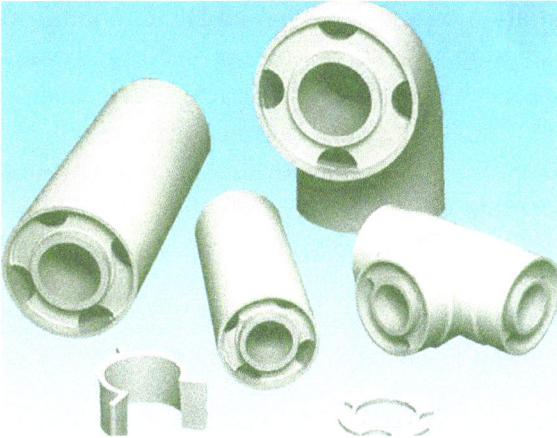

Figure 2 Pre-assembled PP DCP piping systems.

Most pre-assembled systems are limited to the same material for both the carrier and containment piping, such as

- Polyvinyl Chloride PVC x PVC
- Chlorinated Polyvinyl Chloride CPVC x CPVC
- Polypropylene PP x PP
- Polyvinylidene Fluoride PVDF x PVDF

Typical applications for pre-assembled DCP systems are gasoline and jet fuel handling, chemical waste systems, chemical process lines, landfills, and the transporting of other harmful fluids in critical areas.

**Figure 3
Pre-assembled PVC DCP
fitting.**

A variation of a pre-assembled DCP system uses foam-filled insulation instead of spacers in the interstitial areas. This system is not used to contain leaks, but rather to encapsulate the insulation from external corrosive, chemical, and impact damage. A system like this can maintain non-toxic fluid temperatures at a desired level for applications such as hot-water distribution, steam, cryogenics, and condenser and chilled water. Although the initial material cost for this system is high, the real savings comes from eliminating costly external labor-intensive insulation fabrication.

**Figure 4
Pre-assembled copper
insulated piping.**

Custom-Fabricated DCPO Systems

By far the dominant DCP market is for custom systems. In this category, carrier and containment components can be specified in any combination of piping material. Carrier and containment piping combinations consist of plastic and/or metal such as PVC x PVC, PVDF x PP, copper x PVC, steel x PVC, stainless steel x CPVC, etc. The possible piping combinations are almost endless. Applications for these piping systems are the same as those for pre-assembled DCPs.

Retrofit DCP System

Conditions of service in manufacturing plants are continually being revised and changed, as are government codes pertaining to the protection of the environment and the safety of personnel. If codes and/or service conditions require DCP, it is usually simpler and more cost-effective to retrofit an existing single line piping system than it is to replace it with a completely new DCP system. Retrofit systems consist of split piping and fittings used to contain existing single piping systems. Retrofits are currently limited in their containment materials and containment pipe diameters. But, if conditions and budget warrant, it would be prudent to investigate the use of retrofitting existing piping systems.

Ancillary DCP Products

Although dual piping is the heart of the DCP system, there are other critical ancillary products that ensure system integrity. Valve boxes, double-contained tanks and leak detection devises may be required to create a total leak-free system.

Figure 5 PVC x stainless steel DCP waste system for pharmaceutical firm.

Although there are just a limited number of experienced DCP fabricators in North America, their products and systems are proven. For that reason it is recommended that engineers and end-users use the piping systems and resources of these reliable companies rather than—in an attempt to save costs—pay a piping contractor with little or no DCP experience to reinvent the wheel.

As potable water continues to be a critical concern for the public welfare, additional codes and resources will be required to ensure the protection of the billions of gallons in

Figure 6 Clear PVC retrofit system.

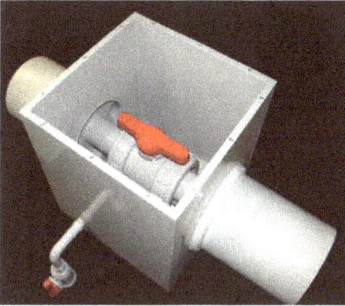

Figure 7 PP DCP valve box.

**Figure 8
PE doubled-contained tank.**

underground water aquifers. OSHA also will continue to insist nothing is more important than protecting plant workers. Clearly DCP systems will be necessary and will certainly play a growing and more important role in delivering efficiency and safety.

**Figure 9 PVC x PVC DCP with visual and electrical
sensor leak detection.**

In applications where leaks are not an option, double
contained piping systems are the best and sometimes the
only option.

Reprinted with permission of the IAPD; issue february/march 2011 – **the
IAPD Magazine**

FACTS REGARDING PVC PIPING
AND THE ENVIRONMENT

Misconceptions about the environmental impact of polyvinyl chloride (PVC) piping systems are widespread, due in part to the controversies which exist regarding PVC in general. However, a careful examination of the facts clearly demonstrates that the negative allegations regarding PVC pipe are erroneous. This article is adapted from the Plastic Pipe and Fittings Association (PPFA) publication: *The Design Guide for PVC Piping Systems for Commercial and Industrial Applications.*

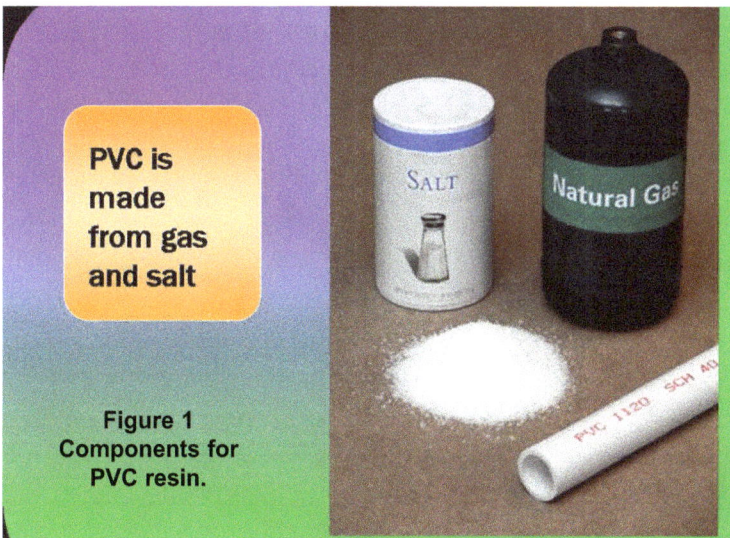

PVC is made from gas and salt

**Figure 1
Components for
PVC resin.**

Resin: By weight, PVC resin is 57% chlorine and 43% hydrocarbon. The chlorine content is derived from salt. The hydrocarbon content is derived from ethylene, which is a derivative of fossil fuel feedstocks. This formulation uses less fossil fuel energy, generates fewer unwanted emissions, and requires fewer fossil fuel resources than other piping material formulations. The abundance and low cost of salt contributes significantly to making PVC more price competitive and sustainable compared to piping materials made of other materials.

Fossil Fuels: Ethylene can be made from either oil, gas, or coal feedstock—offering a wide selection of existing natural resources. A growing percentage of the current production of ethylene is being made from bio-feedstock such as sugar cane. If these new biofuel processing methods prove to be economically and environmentally sound, the elimination of the use of fossil fuels needed to produce PVC could be a reality in the future.

Chlorine: Chlorine is one of the most abundant elements in the world. It is a major component of building materials, packaging, and pharmaceuticals. Used as a disinfectant of water for human consumption, it has saved more lives than any other world health initiative—ever!

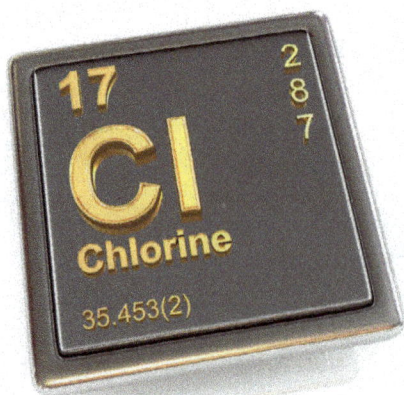

**Figure 2
Chlorine periodic
table element.**

Chlorine is derived from an inexhaustible source: ocean water (and to lesser degree salt mines). In some instances, salt can be derived from seawater using desalination which extracts the sodium chlorine and gives potable water as the resultant end-product.

Durability: The first PVC piping was installed in Germany in the mid-1930s and has remained operational for over 80 years. PVC is immune to electrolytic and galvanic corrosion, scaling, rusting, and pitting. It is also resistant to abrasion, bacteria, fungi, and hundreds of chemicals. Independent studies have shown that municipal PVC piping systems are performance-rated at a minimum of 100 years usage, while concrete and ductile iron piping systems are rated at 85 and 60 years respectively.[1] Increased durability means fewer leaks, better water conservation, and lower costs.

Easy and Safe to Install: PVC piping is one of the easiest and safest piping materials to install. Its advantages include: being light weight, standard 20-foot lengths, simple joining methods, no need for expensive tools, no hot plates or open flames issues, no corrosive protection requirements, minimal or no insulation requirements, and ease of product identification and fabrication.

**Figure 3
A laborer easily carrying
a 12-inch PVC tee.**

Compared to that of other piping materials, the ease and safety of PVC installation reduce on-site accidents and property damage, enable faster project completion rates, and allow installer crews to be easily trained.

Cost Effectiveness: Considering costs across the board—from product, labor, installation, maintenance, and insurance to theft and shipping—PVC piping systems are in most cases the most cost-effective. Costs should be considered when determining the sustainability and utility of any product. Higher upfront and maintenance building costs result ultimately in less investment in new or remodeled construction, higher unemployment rates, and other undesirable factors that negatively affect the well-being of people and communities.

Economic Savings: In a recent study sponsored by the Vinyl Institute and the American Chemistry Council, the use of PVC results in cost savings estimated to be over $9 billion in North America alone, when compared to the costs of using substitute materials.[2] Comparatively, the use of PVC results in savings from lower material and installation costs; savings from less frequent replacement; and savings from less required maintenance and repair requirements.

Recyclable: PVC is totally recyclable. However, because most PVC piping is still in use, not much of it has entered the recycling stream. During PVC pipe production, there is minimal scrap with almost 100 percent of the PVC compound fully utilized. At present, most end-of-life or post-consumer PVC piping is incinerated or sent to land fills. This form of disposition will change in the near future due to the aggressive industry plans to recycle as much PVC products, in all forms, for reuse as possible.

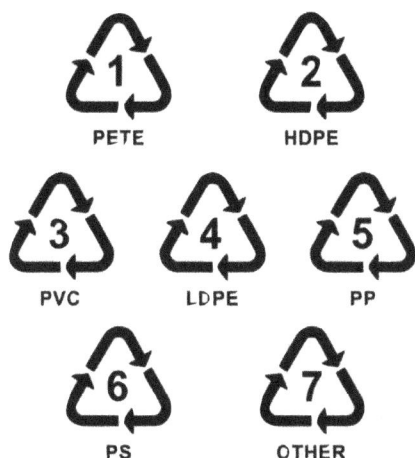

Figure 4
Like most other plastics,
PVC is totally
recyclable.

Joint Integrity: PVC piping can be joined in different ways: flanging, threading, compression couplings, gasketed-bell and spigot, heat fusion, and solvent cementing. The two most commonly used joining methods are solvent cementing and gasketed-bell and spigot. When properly installed, these two methods have working pressures equal to or greater than the pipe or fitting. Studies and history indicate solvent cementing and gasketed-bell and spigot top the list in providing long-lasting joint integrity.

Plasticizers: Rigid PVC piping systems *do not contain plasticizers*. To make PVC material flexible and pliable, plasticizers known as phthalates are added to the final compound. Environmentalists allege that plasticizers are harmful, although there are studies that dispute the charge. However, this issue doesn't affect PVC pipe, which is *rigid* and contains no plasticizers.

Vinyl Chloride Monomer (VCM): In the early 1970s, there were reports that exposure to excessive levels of vinyl chloride in some manufacturing operations resulted in a rare form of liver cancer. In response to this discovery the vinyl industry acted quickly in cooperation with OSHA and the EPA to completely re-engineer vinyl production operations. This action resulted in the elimination of unsafe occupational exposures to VCM and very low emission of VCM to the environment as a result. VCM emissions are now continually monitored and controlled in all PVC processing plants.

Hydrogen Chloride Gas: When PVC is burned, hydrogen chloride gas is emitted. When wood and other building materials burn, lethal carbon monoxide or other toxic gases are emitted. In a typical fire there are much greater amounts of harmful fumes emitted by the burning of wood and other construction materials than by the burning of PVC. Why? PVC piping represents less than 2% by weight in most building construction, whether residential or commercial.[3]

Dioxins: Dioxins are compounds that are suspected of being human carcinogens. Opponents of PVC argue that PVC manufacturing is a major dioxin polluter. This is not true. The largest contributors to dioxin discharge, according to EPA findings, are forest fires, wood-burning fireplaces, coal-fired utilities, metal smelting, diesel trucks, sewage sludge and burning of trash.[4] Studies estimate that the entire PVC industry produces less than 14 grams (less than half of one ounce) of dioxin a year. Another irrefutable fact is that dioxin levels in the USA have decreased 90% in the past three decades, while vinyl production has increased 300% during the same time period.

Life Cycle Assessment (LCA): LCA is a scientific evaluation that analyzes the environmental impact of a material or product from its raw material sources through its production, use, and end-of-life disposition. Several green building rating systems offer points for the use of LCA in the evaluation of competitive materials. European and North American life cycle assessment of PVC piping have shown it to have, in many applications, a favorable environmental impact compared to that of non-plastic piping materials.

Manufacturing Industry Safety: According to 2006 statistics provided by the U.S. Bureau of Labor, the plastic piping industry, from feedstock origination to shipped end-product, has a significantly lower reported incidence of employee illness and injury than other non-plastic piping industries have. In addition, when compared to the average of all U.S. industries, the plastic piping industry has one-third fewer reported employee illness and injury rates.[5]

Figure 5 Injection molding machine used in manufacturing plastics.

Boon to Flora and Fauna: PVC is a material that affords protection to flora and fauna. The use of PVC in siding, decking, fencing, window profiles, pipe, faux auto-panel trim, and other products have prevented the destruction of millions of trees. As a substitute for animal hides and ivory tusks, PVC has prevented the slaughter of thousands of animals.

World Health Issues: The United Nations estimate that more than 6,000 children die each day—that's over two million children a year—due to unsanitary drinking water and waste control. To address these issues, many PVC piping companies and associations have either donated or sold at cost product and engineering services to dozens of non-profits. Because of the superior features and benefits of PVC piping, more and more engineering firms are specifying its use, and by doing so improving the lives of third-world citizens.

Figure 6
Children are our planet's future.

Water Conservation: Experts estimate that 700 water main breaks occur each day across North America, wasting over 2.2 trillion gallons of potable water yearly. The loss of revenues to North American water utility companies total over $3 billion a year. To fix these aging mains, the U.S. government estimates, $23 billion a year for the next 20 years is needed.[6] The American Water Works Association Research Foundation concluded from a survey that the life expectancy rating of PVC surpassed that of any other tested pipe material. A two-year Canadian study found that for each 100 kilometers of water distribution pipe laid, PVC had only 0.7 breaks per year compared to 35.9 breaks for cast iron and 9.5 breaks for ductile iron.[7] Communities around the world are favoring PVC water main and distribution systems due to PVC's documented record of unsurpassed durability and joint integrity.

One Final Word: The obligation to support sustainability touches everyone, in everything we do: the cars we drive, the houses we live in, the vacations we take, the products we buy, even the hobbies that relax us. Today our paramount responsibility is to leave a better planet for the next generation. Now and in the future, the use of PVC piping can help us accomplish our goal of minimizing the human footprint on our environment.

Notes:

1. AWWA Research Foundation, 1998. Quantifying Future Rehabilitation and Replacement Needs of Water Mains.

2. Chlorine Chemistry Division of the American Chemistry Council and the Vinyl Institute, 2008. The Economic Benefits of Polyvinyl Chloride in the United States and Canada.

3. U.S Environmental Protection Agency Municipal and Industrial Solid Waste Division, 1998. Characterization of Building-Related Construction and Demolition Debris in the United States.

4. U.S Environmental Protection Agency (Report No. EPA/600/P-03/002F-2006), 1998. An Inventory of Sources and Environmental Releases of Dioxin-like Compounds in the United States for the years 1987, 1995, and 2000.

5. IAPD Magazine Article-Dec/Jan 2008. "Plastic Piping Systems...Here's to Your Health" Author: David A. Chasis.

6. U.S. Environmental Protection Agency, 2002. Cleanness and Drinking Water Infrastructure Gap Analysis.

7. Infrastructure Laboratory of the Institute for Research in Construction at the National Research Council, 1995. Survey of Water Main Breaks.

Reprinted with permission of the IAPD; issue february/march 2010 – **the IAPD Magazine**

HOMEBUILDERS EMBRACE PEX FOR HOT- AND COLD-WATER DISTRIBUTION

On its own, polyethylene (PE) piping is an excellent product that has many of the valued benefits of other plastic piping systems such as durability; it's easy and safe to install, environmentally sound, and cost-effective. Having the PE polymer chains be linked together by a physical or chemical reaction transforms the piping into a completely different material called PEX (term for polyethylene cross-

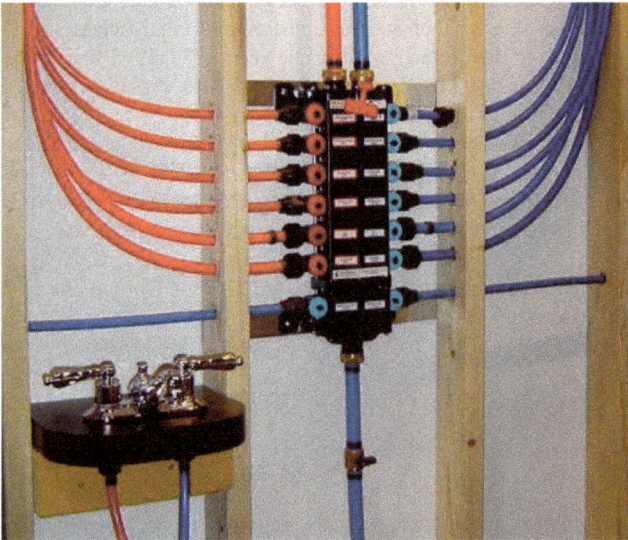

Figure 1 PEX manifold system for hot- / cold-water distribution.

linked—sometime the abbreviation XLPE is also used). The new material is sort of like PE on steroids.

PEX incorporates superior characteristics compared to unlinked PE piping, including increases in: temperature resistance, pressure holding capability, tensile strength, corrosion resistance, creep resistance, abrasion resistance, impact strength, and chemical resistance.

PEX piping was used in Europe in the mid 1970s and was introduced in North America ten years later. There are three distinct methods of manufacturing PEX: Radiation (E-beam), Peroxide (Engel), and Silane. Although the manufacturers use different processing techniques, all PEX piping for use in hot and cold potable water distribution systems in North America must comply with ASTM standard F876/F877 and/or CSA B137.5. Among other requirements, these standards state that PEX piping must adhere to particular temperature and working pressure ranges such as 180°F/82°C @ 100 psi. In addition, PEX piping meets NSF potable water health and safety requirements.

Depending on the application, small tube diameter PEX is available in coil lengths from 100 to 1,200 foot or more, and in colors red, blue, white, and natural. The type of joining connections varies, allowing the installer to select the best joining method for a particular installation as well as crew experience. A contractor can select a system of cop-

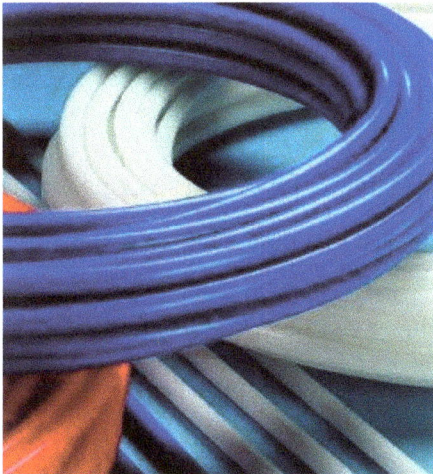

[Figure 2
PEX coils of pipe.

per crimp rings, or stainless steel clamps or caps, with insert fittings of metal or engineered polymer. There is a system of expansion fittings that allow? the PEX piping to shrink around the insert fitting. Last but not least, there is a push-fit system which minimizes labor requirements.

So why specify and replace rigid piping such as copper and steel with PEX? Here are several reasons:

Durable

- More corrosion resistant, allowing tubing to be directly installed in concrete

- Resistant to electrolysis and poor water quality

- Freeze damage resistance minimizes costly ruptures

- Water hammer resistance absorbs hydraulic shock in flexible tubing walls

- Excellent chemical resistance

- Resistance to impact

- Less chance of leakage due to fewer fitting joints required

Easy and Safe to Install

- Can be transitioned to other piping systems

- Lightweight and flexible, allowing ease of use

- Connections made quickly using inexpensive joining tools

- No flames or torches used in installation

- Three installation methods: conventional, remote manifold, or home run

- Tubing coilability simplifies handling and installation
- Color coded to easily identify hot and cold water lines

Environmentally Sound

- Doesn't amplify noise making operating system, more quiet
- Uses less energy in production, shipment, and use
- More sanitary, using no fluxes or lead in joining
- Non toxic and odorless
- Can be recycled into other products
- Life Cycle Assessment is favorable compared to other piping materials

Cost-Effective

- Low purchase price
- Less material price fluctuation
- Less labor to install
- Fewer fittings
- Less on-site theft
- Remote manifold or home run installations cuts costs and uses less energy

With so many features and benefits inherent in PEX piping systems, there are several applications for which PEX is becoming the material of choice: hot- and cold-water distribution systems, hydronic radiant-heating systems (recirculation of hot water in flooring, ramps, driveways), radiant cooling (circulate chilled water instead of hot water), municipal water service pipe, turf conditioning, permafrost protection, reverse osmosis, deionized water, and residential fire suppression systems.

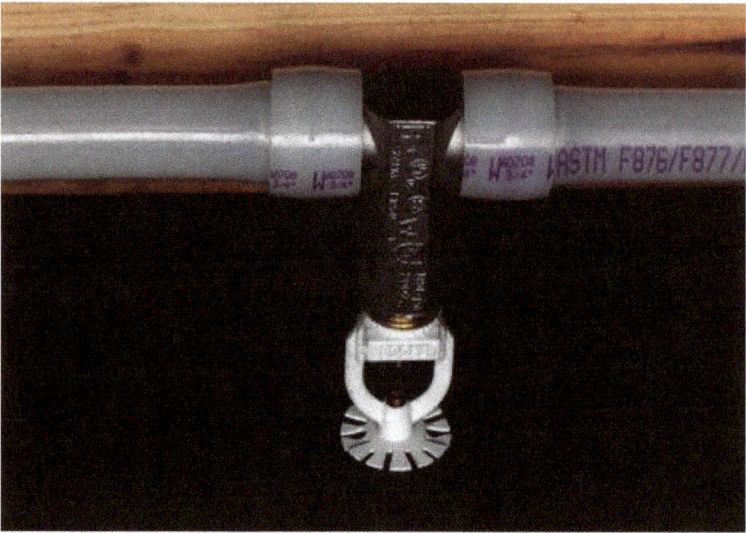

Figure 3 PEX fire sprinkler piping

Figure 4 PEX radiant-heating piping

Next time you are determining what pipe to use for smaller diameter applications, think PEX—one of the fastest growing piping materials in the world.

Reprinted with permission of the IAPD; issue october/november 2011 –
the IAPD magazine

PLASTIC PIPING SYSTEM FABRICATIONS

Thermoplastic piping products of Acrylonitrile Butadiene Styrene (ABS), Chlorinated Polyvinyl Chloride (CPVC), Polyethylene (PE), Polypropylene (PP), Polyvinyl Chloride (PVC), Polyvinylidene Fluoride (PVDF), and other thermoplastics are mostly manufactured from extrusion and injection molding processes. The physical properties of thermoplastics allow these manufacturing processes to produce cost-effective, large quantities of product having exceptional quality. For piping products of small volume and/or unique design requirements, injection molding or extrusion is usually impossible to cost-justify. However, plastic fabrication is an excellent method to produce these special items.

Figure 1 PVC fabricated fitting.

There are several proven methods of fabricating plastic fluid handling products: Hot air/gas welding, bending/belling, solvent welding, heat fusion, machining, and fiberglass reinforcing. The selection of the proper fabrication technique(s) is based on the piping material, design configuration, and the application's conditions of service.

Hot air/gas welding

Thermoplastics can be hand welded using a heat source (welding/extrusion gun), welding rod, and compressed air or nitrogen. Plastic welding, unlike metal welding, produces a material surface bond that normally limits the finished product to 10-psi or less working pressure. Commonly fabricated welded products are: tanks, tank liners, fittings, pipe manifolds, laboratory work stations, fans, blowers, and double-contained piping.

**Figure 2
Skid-mounted
plastic pump and
filtration unit.**

Bending/Belling

By definition, thermoplastics are materials that can be heated, reformed, and cooled with no significant change in mechanical characteristics. This property lends itself to fabricate products that are produced by correctly heating pipe and sheets (usually 4' X 8' in size and of varying thickness) so they become flexible enough to form and maintain a desired shape. Fabricated products using this technique are: duct, hoods, scrubbers, belled or "swedged" pipe ends, pipe bends, tanks, and tank liners

Solvent Welding

Solvent welding of PVC and CPVC (other piping products such as ABS and Styrene are also cementable) is a very common method of making joints. This technique provides the benefits of excellent joint integrity, and is safe and easy to use; in most cases, the joint is stronger that either of the joining products. Fabricated products using solvent cementing are: pipe fittings, pipe manifolds, fans, blowers, flanged products, and many other products.

Heat Fusion

Almost all thermoplastics, when heated to a particular temperature, can be fusion joined. Using butt or socket fusion techniques (not used for most vinyl materials), a permanent plastic joint can be made. With one exception, PVC and CPVC butt-fused joints are not normally pressure rated and are used for drainage applications only. (The exception is a PVC pipe manufacturer that produces a butt-fused joining system for underground use that is pressure rated.) PE, PP, and PVDF butt and socket fusion produces a joint that is equivalent or greater than the working pressure of the products being joined. The benefits of using this method of fab-

Figure 3 PP constructed semiconductor work station.

rication and the products fabricated from this technique are quite similar to solvent welded products. However, unlike solvent cementing, almost all thermoplastic materials can be heat-fused.

Machining

Using specially designed machining equipment for plastics, piping products can be produced using similar metal fabrication techniques. Machined piping products can be fabricated from the alteration of existing products, sheet, and solid-profiles. Commonly fabricated products using this technique are: nuts and bolts, product prototypes, valves and other flow control products, perforations, gauge guards, sight glasses, threading, metal-to-plastic adapters, and flange plates and spacers.

Figure 4 PVC X stainless steel double containment fabrication.

Fiberglass Reinforcing (FR)

In the fabrication process of producing piping products, some of the techniques used do not allow for the finished product to withstand high working pressures. To increase the working pressure of plastic fabrications, FR wrapping is applied to the outer surface of the product. The thickness of the wrap and type of resin used depends on the thermoplastic materials, system working pressure, and chemical makeup of the fluids to be handled. Fiberglass reinforcing is commonly used to increase the pressure ratings for fabricated fittings, manifolds, and tanks.

Figure 5 CPVC fiberglass wrapped manifold.

In many cases, using one or more of the mentioned fabrication methods to produce a product is common. For example, producing a large fabricated fitting for a pressure application with an instrumentation tap could incorporate all of the listed fabrication techniques.

There is an adage in the industry that says, "If you can draw it, an experienced plastic fabricator can make it." There are several experienced plastic fluid handling fabricators in North America. For the name of a component fabricator in your area, contact a local distributor or a piping product manufacturer.

Reprinted with permission of the IAPD; issue february/march 2009 – **the IAPD magazine**

40

PVC...PIPE'S MOST VERSATILE MATERIAL

Plumbing and mechanical engineers who design projects using fluid handling systems have several pressure and drainage piping materials to choose from: steel, cast and ductile iron, copper, concrete, clay, fiberglass, glass, and thermoplastics, including Acrylonitrile Butadiene Styrene (ABS), Chlorinated Polyvinyl Chloride (CPVC), Polyethylene (PE), Cross-linked Polyethylene (PEX), Polypropylene (PP), Polyvinyl Chloride (PVC) and Polyvinylidene Fluoride (PVDF). The question—how to choose the best piping material?

There are several factors that professional engineers use to select piping materials. First, there is the field conditions of service which include factors such as system pressure, fluid temperature, above or below ground installation, chemical compatibility, and type of application. Second, the chosen pipe material must adhere to all applicable building codes. Last, the piping material must provide long-term durability at a cost-effective price—a price that includes material, installation, and maintenance costs.

When conditions of service and codes allow, there is one versatile piping material that seems to stand head and shoulders above most others...PVC.

PVC is a plastic resin made from combining chlorine, extracted from salt (one of the most abundant materials on earth), and ethylene (refined from fossil-fuel feed stocks such as oil and gas). In the future, ethylene may be increasingly sourced from sugar cane or other bio-feedstocks. PVC has been used as a piping material since the mid 1930s. It is the most used piping system in the world because of its durability, safe and ease of installation, cost-effectiveness, light footprint on the environment, and, as we will see, tremendous versatility.

Figure 1
Chemical formula for PVC.

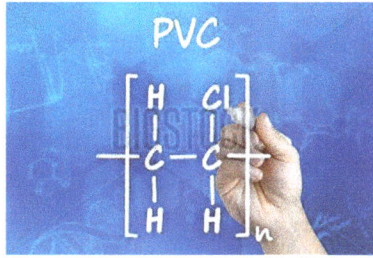

PVC, sometimes referred to as *vinyl*, piping is available in at least eight colors, has seventy distinct pressure and drainage systems, can be easily installed using a minimum of five different joining techniques, has a breadth of product line which is the envy of all of its competitors, and has successfully handled millions of projects in over four dozen different above and below ground application areas.

Breadth of Product Line

The size range of PVC piping begins at 1/8-inch and is available all the way up to 48-inch diameter and larger. Not only is the piping system composed of dozens of sizes and geometries of fittings, it also offers PVC valves, pumps, tanks, duct, scrubbers, fans, hoods, filters, strainers, and other fluid handling products. This breadth of line allows PVC to handle applications where just one piping material is ever in contact with the fluid. But what really distinguishes PVC from other piping materials is its unmatched offering of piping compound and profile systems to handle the many pressure and drainage application requirements. Table 1 list these piping systems including pipe diameter size ranges, pressure ratings where applicable and typical pipe ends.

Table 1 — Applications
Part A: PVC Pressure Piping

PVC Pressure Piping			
Product Description	Pressure Rating (psi)	Nominal Pipe Diameter (inches)	Typical Pipe End
AWWA C900 DR 14	305	4 – 12	gasket
AWWA C900 DR 18	235	4 – 12	gasket
AWWA C900 DR 25	165	4 – 12	gasket
AWWA C900 DR 41	100	4 – 12	gasket
AWWA C900 DR 51	80	4 – 12	gasket
AWWA C905 DR 14	305	14 – 48	gasket
AWWA C905 DR 18	235	14 – 24	gasket
AWWA C905 DR 21	200	14 – 36	gasket
AWWA C905 DR 25	165	14 – 48	gasket
AWWA C905 DR 32.5	125	14 – 48	gasket
AWWA C905 DR 41	100	14 – 48	gasket
AWWA C905 DR 51	80	30 – 48	gasket
Bi-Axially Oriented	AWWA & SDR ratings	4 – 12	gasket
Double Containment	Varies with system	½ – 12	plain
Flexible (swimming pool)	Varies with diameter	½ – 6	plain
Heat Fusible C900	Typical DR ratings	4 – 12	plain
Heat Fusible C905	Typical DR ratings	14 – 36	plain
Heat Fusible Liner	Sch. 40/80 & DR ratings	4 – 36	plain
Horizontal Directional Drilling	305	4 – 6	mechanical
Horizontal Directional Drilling	235	8 – 12	mechanical
P. I. P. SDR 32.5	125	6 – 27	plain/gasket
P. I. P. SDR 41	100	6 – 27	plain/gasket
P. I. P. SDR 52	80	6 – 27	plain/gasket
P. I. P. SDR 64	63	6 – 27	plain/gasket
Reclaimed Water (purple)	Schedule & SDR ratings	½ – 8	plain/gasket
Schedule 120	Varies with diameter	½ – 8	plain
Schedule 80 (gray)	Varies with diameter	1/8 – 24	plain
Schedule 80 (clear)	Varies with diameter	¼ – 6	plain
Schedule 80 (low extractable)	Varies with diameter	½ – 6	plain
Schedule 40 (white)	Varies with diameter	1/8 – 24	plain
Schedule 40 (clear)	Varies with diameter	¼ – 12	plain
Schedule 40 (UV resistant)	Varies with diameter	½ – 6	plain
Schedule 40 (low extractable)	Varies with diameter	½ – 6	plain
Schedule 40 (low flame spread)	Varies with diameter	½ – 2	plain
SDR 13.5	315	½ – 6	plain
SDR 17	250	¾ – 12	plain/gasket
SDR 21	200	¾ – 12	plain/gasket
SDR 26	160	1 – 24	plain/gasket
SDR 32.5	125	1 ¼ – 12	plain/gasket
SDR 41	100	3 – 24	plain/gasket
SDR 64	63	1 ½ – 12	plain/gasket
Well Casing Sch. 40	Varies with diameter	2 – 16	plain/bell
Well Casing SDR 13.5	315	2 – 6	plain/bell

Continued on next page

SDR = standard dimension ratio

Other notes:
1. Pipe diameters and pipe ends vary with each pipe manufacturer's product line
2. Pipe diameter ranges are not static and may change

Table 1 Applications continued

PVC Pressure Piping			
Product Description	Pressure Rating (psi)	Nominal Pipe Diameter (inches)	Typical Pipe End
Well Casing SDR 17	250	2 – 12	plain/bell
Well Casing SDR 26	160	4 – 12	plain/bell
Well Casing SDR 32.5	125	4 – 16	plain/bell
Well Casing SDR 41	100	4 – 12	plain/bell
W. C. Drop Pipe Sch. 80/120	Varies with diameter	1 – 2	threaded

Part B: PVC Drainage Pipe

Product Description	Nominal Pipe Diameter (inches)	Typical Pipe End
Double containment	1½ – 12	plain
Gravity Sewer (D 3034) SDR 23.5	4 – 6	gasket
Gravity Sewer (D 3034) SDR 26	4 – 15	gasket
Gravity Sewer (D 3034) SDR 35	4 – 42	gasket
Gravity Sewer (F 679) PS 46	18 – 36	gasket
Gravity Sewer (F 679) PS 115	18 – 36	gasket
Highway Under drain (F 758) perforated	4 – 8	plain
Schedule 40-Celluar Core DWV	1½ – 12	plain
Schedule 30 DWV	3	plain
Sewer Ribbed (F 794)	8 – 27	gasket
Sewer Corrugated (F 949)	24– 36	gasket
Sewer & Drain-Cellular core PS 50	4 – 6	plain
Sewer & Drain (D 3034) perforated	4 – 6	plain
Sewer & Drain (D 2729) perforated	3 – 6	plain
Storm Drain Ribbed (M304M)	18 – 24	gasket

Part C: PVC Piping/Duct Applications

Other PVC Piping/Duct Applications		
Product Description	Nominal Pipe Diameter (inches)	Typical Pipe End
Central Vacuum – Schedule 20	2	plain
Conduit-electrical	½ – 6	bell
Conduit-utility	1 – 6	bell
Conduit–duct/phone	2 – 4	bell
Duct (air & fumes)	6 – 24	plain
Flue Gas Venting- Schedule 40	1½ – 4	plain
Radon Elimination -Schedule 40	2 – 4	plain

Notes:

Acronyms:
AWWA = American Water Works Association
DR = dimension ratio
P.I.P. = plastic irrigation pipe
PS = pipe stiffness (value relates to impact resistance)

Variety of Colors

Color is added in PVC compounds before processing and theoretically can produce any color imaginable. The most common PVC piping system colors and applications are shown in Table 2.

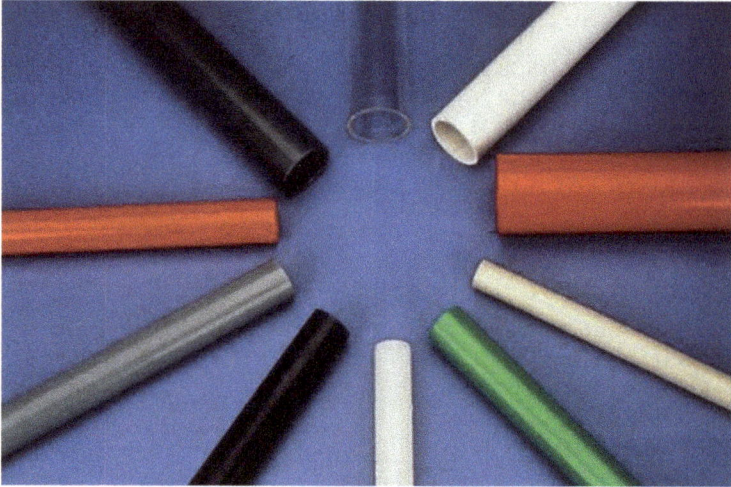

Figure 2   Variety of colored PVC piping.

Joining Methods

The two most commonly used methods of joining PVC piping are solvent cementing and gasketed-bell spigot joints. Most above ground PVC applications are solvent cemented. Although belled-gasketed joining is the most popular method of joining PVC for buried applications, solvent cementing and heat fusion are also used as well. Other methods of joining PVC piping systems are: flanging, threading, mechanical grooving, and push-fit. Table 3 lists the various PVC joining methods with comments on their use.

Table 2 – PVC Color Piping Guides by Applications*

PVC Piping Color Guide by Applications*	
Color	**Common Applications**
Blue	Water main, high purity & mining
Brown	UV resistant
Clear	Industrial, sight gages, UV resistant
Gray	Industrial, duct and conduit
Green	Sewer and drain
Purple	Reclaimed, recycled and/or non-potable water
Translucent	Low flame and smoke containment
White	DWV, well casing, irrigation, pressure, and most others
Yellow	Mining and dewatering

*Note: The colors listed are a guide. Always check your local codes and the pipe markings to ensure the pipe is rated for the application. Other colors are available especially for applications like underground ducting.

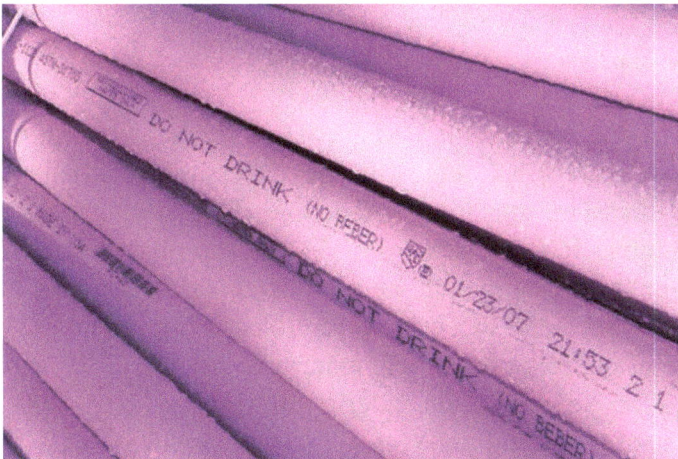

Figure 3 Purple PVC pipe for reclaimed non-potable water.

TABLE 3 – Most Common Joining Methods for PVC Piping Systems

Joining Method	Comments on Use
Solvent Cementing	Above or belowground use—more frequently used for joining aboveground installations—no special tools required—joints need to cure before testing and use—one of the strongest and best joint integrity of any PVC joining method
Gasketed Bell	Underground use—quick and simple to install— needs thrust blocking—no special tools required—excellent track record of joint integrity
Flanging	Aboveground use—150-psi working pressure for most flanges could limit system pressure—easy to install—can connect and reconnect piping—transition method from PVC to other piping materials—good choice for large diameter piping—can prefabricate system components for shipment to job site
Threading	Above ground use—direct threading of Schedule 80 or 120 only—direct threading reduces working pressure 50%—thread by socket cemented adapters are available to handle working pressures to 150°F and to join dissimilar piping materials
Mechanical Grooving	Above ground use—easy to install—rolled or cut grooved pipe—may reduce working pressure of piping systems—elastomerically sealed—good choice for large diameter piping—can prefabricate system components for shipment to job site
Heat Fusing	Used in piping fabrications—new system with special compound can be heat fused in the field for underground and lining applications—need sophisticated and expensive joining equipment and specially trained installers
Push-Connect	Above ground use—for pipe diameters 2-inches and below—metal grabber may be exposed in transporting the fluid—very easy to install

Applications

If the temperature range of the piping application is between 32°F/0°C and 140°F/60°C, less than 250-psi working pressure, and fluid compatible, PVC may provide the solution. Table 4 lists over 50 applications in residential, commercial, municipal, and industrial markets where PVC piping has been successfully installed and in many cases is one of the preferred piping materials.

Can you possibly understand why anyone would not consider PVC piping systems if the conditions of service and codes allow? This article considers PVC only for bona fide piping applications. How about the many unintended

Table 4 — Common PVC Piping Applications

Residential	Central Vacuum	Septic System
	Drain/Waste/Vent	Sewer
	Irrigation	Swimming Pool
	Radon Removal	Water Distribution
	Rain Harvesting	Well Casing
	Reclaimed Water	

Commercial	Building Service	Reclaimed Water
	Lines	Refrigeration
	Chilled Water	Sewer
	Condensate Drain	Site Utilities
	Condenser Water	Swimming Pool
	Drain/Waste/Vent	Water Distribution
	Irrigation	Water Towers
	Rain Harvesting	Well Casing

Municipal	Aquariums	Rain Harvesting
	Amusement/	Slip Lining
	Water Parks	Sewer and Drain
	Bridge/Road	Swimming Pool
	Drainage	Water Distribution
	De-watering	Water Mains
	Landfills	Water Towers
	Irrigation	Water/Waste Treatment
		Well Casing

Industrial	Air Pollution	High-purity Water
	Systems	HVAC
	Chemical	Mining
	Processing	Marine
	Desalination	Pharmaceutical
	De-watering	Pulp/Paper
	Double-	Refrigeration
	containment	Semiconductor
	Drain/Waste/Vent	Steel Making
	Ducting	Surface Finishing
	Electronics	Water Distribution
	Fire-hydrant	Water/Waste Treatment
	Fish Farms	
	Food/Beverage	

Figure 4 Double containment PVC pipe with stainless steel pipe.

applications PVC piping has been used for, such as: pipe furniture, shipping containers, shelving for stacking wine or building plans, pick-up truck containers for carrying smaller diameter pipe, "Blue-Man Group" musical instruments, playground equipment, concrete forms, plant water-containers, and horserace track railings to name a few.

When PVC is being considered for a piping project, keep in mind all of the reasons why it is the post popular piping system on the planet: durable, safe and easy to install, environmentally sound, cost-effective and versatile. PVC: The preferred piping material.

Reprinted with permission of the IAPD; issue august/september 2011 – **the IAPD magazine**

RECYCLING PVC PIPING SYSTEMS

Recycling of all manufactured products is getting a lot of press for being the environmental way of the future: rightfully so. But for a product that is resistant to chemical and corrosive attack and could have a service life well over a hundred years without significant decline in performance, you would think the environmental activists would tone down their rhetoric somewhat regarding PVC piping systems—systems that are matchless when it comes to being leak-proof, durable, easy and safe to install, cost-effective, and environmentally sensible. These attributes have been repeatedly proven in long-term performance results when compared to other non-plastic piping materials. So, why all the hoopla about recycling PVC pipe?

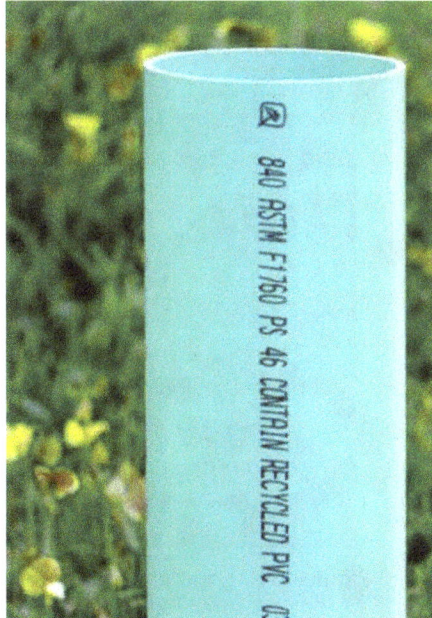

Figure 1 Drainage pipe using recycled PVC (not for potable water)

Most likely the cause that provokes the fury of the environmental activist is the arena of short-term product life (ten years or less) of such items as packaging, electronics, clothing, and automobiles which tend to spotlight concerns for all manufactured products no matter what the life of the product may be. But to be fair, let's examine the recycling of PVC pipe and see if we can ascertain the economic, energy use, and environmental impact it has on our planet.

In recycling terms, PI does not stand for "Private Investigator" but rather "Post Industrial," which is the name given for scrap left over from the manufacturing process. In the manufacture of PVC piping, PI includes mostly scraps developed from production start-up, shut-down, out-of-spec pipe and quality control testing samples. This scrap is routinely reground and later introduced back into the production process. It is estimated that less than 1 % of the PVC material used by piping and fitting manufacturers is not used to produce product. In other words, PVC pipe manufacturing has excellent marks in PI recycling.

Post Consumer (PC) recycling reclaims material after the product has been used for its intended purpose. When PVC

Figure 2 Regrind at a PVC pipe plant for inclusion into making new pipe.

pipe and building products such as siding, window profiles, fencing, or decking are discarded due to remodeling, new construction debris, demolition, or piping relocations, you would think the manufacturers could recycle these materials. Unfortunately, it is not as easy as it sounds, and here's why.

Compounds and methods to produce *extruded* PVC pipe are completely different for fittings and other PVC *molded* products. Every manufacturer can have a slightly different recipe than its competition for each product line. However, all compounds must conform to ASTM, NSF, and PPI standards. Therefore, to recycle PVC piping, the product components must be returned to the original manufacturer.

This action could be hindered in that most piping systems consist of components from multiple manufacturers because most PVC piping products are interchangeable. Plus, what do you do with solvent-cemented pipe and fitting joints that are basically impossible to take apart? To insure conformance of dimensions, burst, and working pressure, the ASTM standards limit the amount of recycled material that can be mixed with virgin material to produce pressurized pipe and fittings.

In addition, these piping systems may be contaminated from previous use that would not meet NSF requirements to transport potable water. It is possible that ASTM and NSF standards may be waived if the manufacturer uses post consumer material in the manufacturing of non-pressure pipe or fittings for non-potable water applications.

"Take-back" programs of PVC piping products are presently being discussed by industry manufacturers and their associations in order to lessen the impact on the environment. Although the take-back goods may be limited in piping products, they can certainly be reground and used in creating a myriad of products that incorporate the lightweight, chemically resistant, and maintenance-free benefits of PVC.

Today, most post-consumer PVC piping products are incinerated or put into land fills. It's so "un-green," you say. That's true to a point, but there is an extremely small amount of PVC piping products when compared by weight to other building products not recycled. A study done for the U.S. Environmental Protection Agency determined that 49% of new construction and residential and commercial demolition debris was actually recycled. Most of this debris consisted of concrete, wood, metal, and roofing materials. PVC piping represented *0.4%* of all the measurable debris.

Figure 3 Recycled PVC used to make a ruler.

It is true that PVC piping in land fills do not decompose. In fact, it is one of the primary reasons that PVC pond liners and piping to vent methane is used in landfill sites. But again, the long life and small amount of PVC piping being disposed produces a rather insignificant impact to the environment. In addition, PVC can be safely incinerated in modern facilities with its energy recaptured and used while contributing no damaging emissions.

Yes, we should be concerned about recycling PVC piping systems. But also realize that PVC piping is one of the most durable products on earth with arguably no known end-life in some particular applications. Its outstanding durability will have PVC piping systems outlasting many of the structures or underground applications in which it is installed. Viva le PVC!

Reprinted with permission of the IAPD; issue december 2007/january 2008 – **the IAPD magazine**

SUPER PIPE

Conventional Polyvinyl Chloride (PVC) pipe has been successfully installed for over three-quarters of a century and has become one of the most abundant piping systems on the planet. Why? It is durable, easy and safe to install, cost-effective, and environmentally sound. So how can you improve upon a product that for so long has had market share in dozens of applications? The answer is—bi-axially molecularly oriented PVC pipe, or PVCO for short.

When conventional PVC pipe is extruded, it is made by a fully automated, one-step, continuous extrusion process that results in the final product having its vinyl polymer chains oriented randomly. PVCO pipe undergoes a controlled stretching process after extrusion, which alters the pipe material by orienting the polymer chains in a specific and ordered fashion.

Molecularly oriented pipe is not a new concept. The method for manufacturing PVCO was developed in the mid seventies in Europe and was comprised of a two-step or "batch" process. Conventional PVC pipe was extruded, and then each pipe length was expanded under controlled temperature and pressure in a large mold. Although this process was effective, it was difficult to control and automate.

Most PVCO pipe is now made using a continuous process by stretching a thick "feed stock" PVC profile over a precisely sized mandrel under tightly controlled temperatures and pressures. This process results in creating PVC pipe with its polymer chains orientated in two directions, instead of randomly—in both a circumferential and longitudinal direction—hence the industry's label of "bi-axial oriented" pipe.

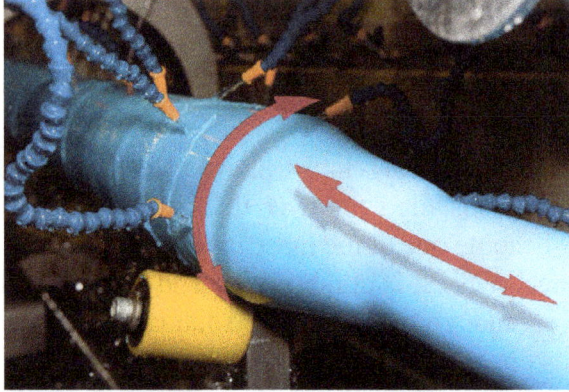

Figure 1 Orienting PVC piping axially and circumferentially.

Improvements to the original orientation process have been made in the 1990s by European developers who have met the challenge of producing a quality and price competitive PVCO piping product using an automatic in-line process. An important side benefit of this automated process, besides lower production costs, is that PVCO pipe is available with closely controlled external and internal pipe diameters, making pipe wall tolerances even tighter than conventional PVC.

Improved Features

Ok, so why the excitement over molecularly oriented pipe? The answer is that PVCO pipe maintains the many qualities of conventional PVC pipe such as excellent chemical resistance, non-corrosive, ease of installation, durability, low thermal conductivity, smooth flow, fully recyclable, low or no maintenance requirements, code compliant, resist-

Figure 02 In-line process for manufacturing PVCO pipe.

ant to water permeation, and cost effective while having several major product improvements. These improved features are:

Hoop strength

PVCO pipe has almost double the Hydrostatic Design Basis (HDB) of PVC (7100 psi versus 4000 psi). [HDB is used with a given service factor to calculate Hydrostatic Design Stress (HDS). HDS is used as the basis of calculating various plastic pipe pressure ratings.] This feature allows thinner-walled pipe, larger internal diameters, and reduced transport energy costs compared to similar diameter and pressure-rated PVC piping.

Lightweight

PVC pipe is lighter in weight than most piping materials; however, most PVCO pipe is on average 40% lighter than PVC. The lighter weight results in an easier-to-install and less expensive pipe installation. For example, two men can easily carry a 20-foot length of 12-inch diameter PVCO pipe.

Figure 3 Man carrying a 20-foot length of 10-inch diameter PVCO pipe.

Impact resistance

The impact strength of PVCO pipe is 3 to 5 times that of PVC. This feature results in a much tougher and resilient product that can withstand more field site abusive conditions than other piping systems, even at temperatures as low as 32°F/0°C.

Resists cracking

PVCO's pipe wall is composed of stratified material versus a single layer pipe wall in PVC. This unique feature increases notch resistance, thereby minimizing splitting and crack propagation. This characteristic also adds to the safety and ease of tapping into pipe walls in underground applications and bodes well for future grooved-piping above ground uses.

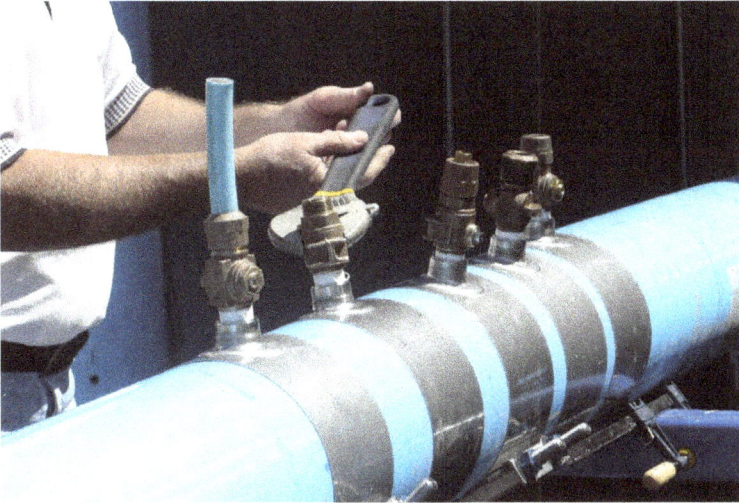

**Figure 4 Tapping PVCO pipe is quicker and stronger
than PVC pipe.**

Flexible

Like impact resistance, pipe flexibly is an extremely important characteristic for heavily trafficked areas. And, having lower bend radii than PVC may mean fewer fittings are required, resulting in lower installed costs.

Cyclic fatigue resistance

Force main and irrigation piping systems normally are subjected to many cycles of flow throughout their use. With much higher hoop stress capability than conventional PVC, PVCO is an excellent choice for increasing the piping life of cyclic applications over most all competing materials.

Water hammer

The lower pressure-wave velocities of PVCO piping compared to other piping systems reduces or eliminates the possibility of pipe damage due to excessive surge pressure or water hammer.

Figure 5 PVCO pipe will flex and is difficult to break.

Environmentally sound

Due to PVCO pipe's thinner wall and reduction in pipe weight, there is much less embodied energy in PVCO than other piping materials. Plus, the thinner walled pipe increases the inner pipe diameter to optimize fluid flow while reducing the amount of energy to transport fluids.

Figure 6 Toughness of PVCO pipe to withstand crushing.

Conclusion

Are there limitations for PVCO piping systems? Sure! Presently, almost all PVCO applications are used for underground pressurized systems such as water mains, sewer force mains, irrigation systems, and industrial lines. Pipe diameters are limited from 4-inch to 16-inch in cast iron and iron pipe dimensions. The typical joining method used with PVCO piping is gasketed-bell. While research has shown that solvent cementing can be used to join PVCO pipe, it is currently not supported in existing AWWA and ASTM standards. It is very likely that on-going research will shortly produce a reliable and certified solvent welded pipe joining system for PVCO. One other concern to note is the process of making molecularly oriented pipe prevents heat fusion techniques to be used in joining pipe or fabricating fittings.

In the future, except for drain-waste-vent and other low-pressure applications where excessively thin pipe walls could present problems (especially in above ground installations), PVCO will become more and more prevalent. The most likely expansion of PVCO pipe in the near future will be in AWWA water main applications where pipe process technology will soon be able to increase the pipe diameter offerings to 24-inch or higher from the present 16-inch maximum. Once a compatible solvent welded system is developed and proven, many above ground commercial and industrial applications could be available for PVCO piping use.

There has not been a more significant development in the North American piping industry in the last 10 years than the advent of biaxially oriented PVC. Whether one considers quality, installed costs or an environmentally light footprint, no other piping system can quite compare to Super Pipe—PVCO.

Reprinted with permission of the IAPD; issue august/september 2009 – **the IAPD magazine**

THERMOPLASTIC FLANGES

Flanging is the system of choice when joining thermo-plastics to dissimilar piping systems, especially larger diameter piping systems. In addition, flanging is the third most popular industrial thermoplastic piping joining system, after solvent welding (cementing) and heat fusion systems.

Figure 1 Spigot flange.

There are several inherent advantages of using a flanged piping system.

Can disassemble and assemble. With solvent welding and heat fusion joints, once the joint is made, the only way to "unmake" the joint is to cut the joint out of the system and start again. Flanged systems can be assembled and disassembled infinitely.

Easy to fix leaks. If there is a leaking flanged joint, tightening up the flange bolts or replacing a defective gasket handles 95 percent of field problems. Also, a flanged system

278

simplifies isolating a piping system for maintenance or field repairs.

Can join dissimilar piping materials. With standard flanges adhering to ANSI/ASME B16.5 bolt-patterns and dimensions, thermoplastic piping systems are easily flanged to other piping systems.

Can prefabricate a system for field installation. When field conditions are a concern and/or skilled installers are scarce, prefabricating a flanged system in a controlled environment and with skilled workers can be a boon for many field installations.

Ideal for large diameter pipe sizes. When installing piping systems 8-inches and larger in diameter, it would be advisable to consider a flanged system due to the ease of installation compared to solvent welding and heat fusion joining. Flanging is also extremely useful when installing complex piping systems in confined spaces in which positioning cumbersome joining tools are a hassle or impossible.

Figure 2 Blind flange.

The concerns in using a flanged system.

Possible corrosive issues. Gaskets and bolting hardware are of different material composition than thermoplastics and could be susceptible to chemical attack.

Working pressure limitations. Flanged systems are limited mostly to 150-psi working pressure ratings (more on this subject later).

For above ground applications only. This joining system cannot be directly buried in earth bearing soils due to point-loading of the flange connections.

Could have space limitations. Flanging is dimensionally less compact than other piping systems.

Figure 3 Socket flange.

There are three types of flanges for use in joining thermoplastic piping systems—solid one-piece, van-stone, and blind. Most flanges up to 12-inch in diameter are injection molded. Flanges larger than 12-inch are usually van-stone types made using molded and fabricated parts. Except for blind flanges, the flange hub end-connections are configured as either socket or spigot types. The spigot flange is designed for butt-fusion (pipe-end to pipe-end connection) or for insertion into a fitting socket (very useful when flanging valves or fittings). The socket flange is designed for joining using solvent cementing or socket heat-fusion techniques. Van-stone flanges are of a two-piece design in which the bolting ring is allowed to rotate independently of the hub, greatly simplifying the bolting process. Usually there are two ring materials to choose from depending on the application: plastic or metal. Blind flanges "dead-end" or cap a piping system and are removable for future piping continuations.

Figure 4 Van stone flange.

As mentioned earlier, thermoplastic flanges, like other ASTM F1970 designated products such as unions and valves, are limited in most cases to a 150-psi working pressure rating at ambient temperature by flange manufacturers. In the last two decades, several manufacturer of unions and ball valves, by redesigning their products, have increased the pressure ratings of their product to 235-psi or 16 bars (1 bar = 1 atmosphere = 14.7 psi). With the increase in pressure capabilities of unions and valves, thermoplastic flanges could be the limiting factor in establishing maximum working pressure of any piping system in which they are installed.

There are a few thermoplastic flange manufacturers who offer both 150-psi and 300-psi bolt patterns. The 300-psi flanges can handle greater working pressures and are normally fabricated from sheet and/or bar stock. These products may have longer delivery times than standard 150-psi flanges.

Figure 5 Bolted plastic flange.

Thermoplastic flanging joining techniques are similar to other piping systems. The salient points to remember when flanging plastic flanges are:

- Use a torque wrench and follow manufacturers' bolting pattern and torque requirements when tightening nuts and bolts.

- Heavy flat washers should be used on flanges to be joined.

- Use lubricated bolts (lubricant to be material-compatible).

- Do not use ring gaskets; use chemically compatible full-face gaskets with a 55—80 durometer.

- Avoid joining plastic flanges to metal flanges with excessively raised inner lip.

Flanging will continue to be used extensively in thermoplastic piping systems and, when conditions demand, will often be the joining system of choice.

Reprinted with permission of the IAPD; issue august/september 2004- **the IAPD magazine**

THE ULTIMATE PIPING MATERIAL

Plastics are the fastest growing and one of the most popular piping materials in the world. Polyvinyl Chloride (PVC) is the dominant material with over two-thirds market share followed by the Polyolefins—Polyethylene, Polypropylene, and Polybutylene. However, for those special hard-to-handle applications, few piping systems can match the performance of Fluoropolymers.

Figure 1 PVDF natural, pigmented, and lined-steel pipe and fittings.

In the mid-1960s, chlorinated plastics—PVC and Chlorinated Polyvinyl Chloride (CPVC)—were just starting to hit their stride. Yet, for temperatures above 210°F/99°C and/or chemical resistance to many inorganic and organic fluids, no plastic pipe was readily available. For a short period of time, there was a chlorinated polyether material, Penton®, which could handle higher temperatures and had better chemical resistant properties than vinyls. However, due to difficulties experienced in manufacturing, installation, and high costs, Penton® never dented this special engineered portion of the piping market. In the later part of the 1960s, there were new materials that entered the piping market—fluorinated plastics.

Figure 2 PVDF valves.

Several engineered fluoropolymer piping compounds were introduced over four decades ago— Ethylene-chlorotrifluroethylene (E-CTFE), Perfluoroalkoxy (PFA), Polytetrafluoroethylene (PTFE), and Polyvinylidene Fluoride (PVDF). Other fluoropolymers were also introduced—Polychlorotrifluoroethylene (PCTFE), Ethylenetetrafluoroethylene (ETFE) and Fluorinated ethylenepropylene (FEP)—but they were rarely used in solid-wall piping applications. All of these compounds are hydrophobic (lack of affinity for water absorption), can handle temperatures exceeding 280°F/138°C (PFA and PTFE can be used in some applications exceeding 400°F/204°C), have excellent chemical resistant properties, and have very low fire and smoke development indices.

However, due to physical properties, durability, ease of use, and cost-effectiveness, two fluorinated solid-wall piping compounds have made the most impact in the market place—E-CTFE and PVDF.

E-CTFE (Halar®) and PVDF (Kynar® and Solef®) have emerged as the two most popular rigid solid-wall piping systems in most part due to their cost effectiveness and chemical resistance to highly troublesome aggressive fluids. In addition, both of these piping compounds have excellent physical properties such as excellent abrasive resistance, hardness, impact strength, burning rates, brittleness resistance at low temperature, and high mechanical strength.

Figure 3 PVDF ultra high purity water piping system.

With these inherent properties, E-CTFE and PVDF have made inroads into the chemical processing industry, semiconductor industry requiring very low extractable values, pulp and paper industry in handling halogens and acids, nuclear waste processing in handling radiation and hot acids, water and waste treatment plants for handling high concentrations of sodium hypochlorite and membrane filtering systems, pharmaceutical and food and beverage industries requiring low extractables, and FDA-approved compounds. And they are ideal piping materials for use in air plenums due to their low fire spread and smoke development. Although these piping compounds can be pigmented (added color) or natural, non-pigmented is normally the material of choice where low extractables are required.

Figure 4 PVDF chemical-waste systems.

When do you choose PVDF or E-CTFE? Assuming both materials can handle the conditions of service for a particular application, it seems that PVDF has an edge over E-CTFE for several reasons. First, there is the breadth of product line. PVDF offers complete piping and ducting systems in pipe diameter sizes from ½-inch to 12-inch whereas E-CTFE piping systems are normally limited to 4-inch diameter and below. E-CTFE piping systems are also limited in its offering of valves, flow monitoring devices, filters, strainers, pumps, and other fluid handling products compared to PVDF.

Second, there are more manufacturers of PVDF compound and processed products than of E-CTFE, keeping pricing and product availability more price and delivery competitive.

Third, PVDF is much more publicized and specified in the engineering community through the decades of marketing efforts of the compound and product manufacturers. For example, PVDF is the only fluoropolymer being offered as a complete chemical waste drainage system.

Fourth, PVDF is more cost effective than E-CTFE in most applications due to its much higher tensile strength, which allows thinner-walled PVDF piping to have the same working pressure ratings than thicker-walled E-CTFE pipe.

There are conditions of service where chemical resistance of hard-to-handle fluids such as hot amines or molten alkali metals and/or continuous temperatures above 300°F/149°C can only be handled by the Teflon® materials—PTFE and PFA. Both of these materials have limited solid-wall piping lines, but are extensively used as liners for flanged-steel pipe in a variety of pipe diameters (PVDF and E-CTFE are also available as liners). Due to the high cost of these compounds (on a per pound-product basis) and/or the special installation techniques required, it is safe to say that

Figure 5 PVDF mechanically joined system in pharmaceutical plant.

the Teflon® piping system usage is to be explored usually as the last resort (and maybe the only resort) in selecting a piping systems for extreme conditions of service.

There are few piping systems having the features and benefits to better handle aggressive fluids and/or fluid temperatures above 280°F/138°C in a durable, safe, and cost-effective manner than fluoropolymers—the ultimate piping material.

Reprinted with permission of the IAPD; issue april/may 2009 - **the IAPD magazine**

WHY PVC PIPING SYSTEMS?

Plastics are the fastest growing piping systems in the world and have been for over the last four decades. Why? They are environmentally sound, easy and safe to install, reliable, long-lasting, and cost-effective. Today, plastic piping is one of the preferred materials in water mains, sewer lines, irrigation systems, swimming pools, drain-waste-vent lines, water well-casings, chemical and acid drainage systems, natural gas distribution lines, and slip-liners for dam-

Figure 1 PVC lightweight allows the ease in carrying a 12-inch diameter fitting.

aged municipal piping systems to name a few. Of all the plastic pipe materials in use, Polyvinyl Chloride (PVC) is by far the most popular (estimated PVC usage is over 70% of footage for all plastic piping installations). Why?

PVC resin is made (by weight) from 43% petroleum-based feedstock and 57% salt. This means PVC production uses less energy, generates fewer emissions, and requires fewer natural resources than many other piping materials. Because salt is a relatively inexpensive compound and one of the most voluminous materials on earth, PVC is more sustainable and price competitive compared to other piping materials.

Another benefit of PVC is it has one of the highest strength-to-weight ratios of any commonly used polymer piping materials. This feature allows PVC to have a lesser pipe wall thickness to maintain a designed working pressure compared to other materials. Two other valuable physical characteristics of PVC are the high modulus of elasticity and low coefficient of thermal expansion compared to other non-metallic materials. These properties minimize the amount of expansion and contraction of PVC piping sys-tems, allowing reduced piping offsets, loops, and expansion joints.

**Figure 2
Metal valves
corrode—
plastic valves don't.**

When it comes to installing PVC piping systems, no other piping system incorporates the variety of available joining techniques. By far the most popular and least costly way of joining PVC to PVC in a leak-proof system is by solvent welding (cementing). This joining method has been used for over 60 years and, if done properly, produces a non-leaking, homogenous, monolithic joint that has a greater pressure rating than either the pipe or fitting. This system requires no fancy or expensive tools either. Other successfully used joining systems for PVC piping are: flanging, threading, gasketed-bell and spigot, rolled or grooved o-ring compression, transition adapters, and heat fusion.

Figure 3 PVC fabricated fittings.

Not only does PVC have a variety of joining methods, but the breadth of product line is unmatched. Pipe diameters start at 1/8-inch and are available up to 24-inch and larger. PVC is available in pressure ratings of Schedule 120, 80, and 40 as well as several constant pressure ratings of 250, 200, 160, 100, and 50-psi, and ducting. PVC piping is also available in a multitude of colors, lengths, pipe joining ends, and specially formulated compounds. These compounds can increase impact resistance, add additional ultraviolet protection, produce a transparent pipe, or add resistance to the extraction of contaminants for high purity water piping systems.

Figure 4 PVC piping used in municipal pool pump and filter complex.

By incorporating the above listed benefits and all the normal cost saving advantages of plastic piping—light weight, optimum flow rates, excellent abrasion resistance, minimum maintenance, outstanding chemical and corrosion resistance, low thermal conductivity, durability, flexibility, reduced freight costs, and reduced job-site theft—PVC, when conditions of service are applicable, is unquestionably one of the most reliable and cost-effective piping materials ever created.

Reprinted with permission of the IAPD; issue june/july 2008 – **the IAPD magazine**

WHY PLASTIC VALVES?

You know what really irks us plastic piping zealots? Seeing residential, commercial, and industrial plastic piping system installations using metal valves. Many engineers may not even know that plastic valves exist although they specify plastic piping in most parts due to its proven durability, ease and safety of installation, environmentally soundness, and cost-effectiveness. It's estimated that plastics can handle 70% or more of piping applications that fall in thermoplastics usage conditions of service (not exceeding 250°F/121°C or 230 psi); this means engineers have a multitude of opportunities to specify plastic valves.

Figure 1 PVDF True Union Ball Valve.

Table 1
Commonly Used Plastic Valve Availability

Types	Materials and Diameter Size Ranges (Inches)			
	CPVC	**PP**	**PVC**	**PVDF**
Angle	¼–½	¼–½	¼–2	¼
Ball	½–4	½–4	½–4	½–4
Butterfly	1½–12	1½–24	1½–12	1½–12
Check	½–12	½–8	½–12	½–8
Diaphragm	½–6	½–10	½–10	½–10
Float	N/A	¼–1	N/A	¼–1
Foot	½	½–4	½–4	½–4
Gate	½–8	½–14	½–14	N/A
Globe	N/A	½–4	½–4	½–4
Goose Neck	¼–½	¼–½	¼–½	¼–½
Laboratory	¼–3/8	N/A	¼–3/8	N/A
Multiport	½–4	½–4	½–4	½–4
Needle	¼–½	¼–½	¼–½	¼–½
Pressure Relief	½–4 ½–4	½–4	½–4	½–4
Pressure Regulator	N/A ½–3	½–3	½–3	½–3
Solenoid	½–3 ½–3	½–3	½–3	½–3

N/A = not available

By definition, a valve is a device that regulates the flow of a fluid (that includes gases, slurries, and liquids). There are well over a dozen valve types, each regulating flow in a precise manner to offer maximum performance at the lowest cost. Table 1 lists most of the commonly used valve types, all of which are available in many of the standard thermoplastic piping materials and size ranges shown (due to the dynamic nature of the plastic valve industry, material and diameter size shown may have changed).

So why else should an engineer switch from being a valve *metal-head* to a valve *plastic-head*? Durability first comes to mind. By heat fusing, solvent welding, or flanging a plastic valve in a plastic piping system, joint integrity is enhanced; galvanic and electrolytic corrosion are eliminated. Plus, there is not a chance of dezincification (removal of zinc from a metal alloy—usually brass—due to attack from aggressive fluids) which could lead to valve failure. In addition, when a transported fluid has wetted contact with only one compatible piping system material (plastic), there are no hidden concerns regarding the chemical compatibility of other piping materials. Other features that foster system durability in plastic valve construction are: increased abrasion resistance; greater permeability resistance of transported fluids; and no pitting, rust, or scaling.

Figure 2
PVC solenoid valve.

Many engineers may be not as concerned as installers to the labor saving benefits of installing plastic valves; however, these benefits are real and make for a safer and easier installation. In most cases, the plastic valve's light weight (an average of 1/16 the weight of a metal valve) reduce or eliminate the need for heavy moving equipment as well as heavy and bulky valve supports. When installing smaller diameter valves, many installers will use threaded or flanged connections to join metal valves to plastics, which could create potential leaks due to the expansion and contraction differences of dissimilar materials. Also, due to the compressive strength differences between metals and plastics, over-tightening of metal threaded or flanged end-connections to plastics can possibly cause cracking of the plastic products.

Figure 3 PVDF globe valve.

What about the environmental soundness of plastic compared to metal valves? First, there's the lead issue. With plastic valves, lead level concerns are a moot point because lead is not present in any of the plastic valve materials. Many metal valves, especially those in residential applications, are supplied by off-shore manufacturers who may not diligently monitor the acceptable lead levels of their product for the North American market. Second, due to their smooth walls and lack of corrosive buildup, plastic values reduce friction loss; hence, they reduce the energy required to transport fluids. Another environmentally sound benefit of plastic valves is based on preliminary Life Cycle Assessments—a scientific method used by many building environmental rating agencies to determine the environmental impact of a product from cradle to grave. Findings show that plastic piping products seem to have a more favorable impact on the planet than many other piping products.

Figure 4
PVC Butterfly valve.

A sensible method to determine the cost effectiveness of any piping system is to examine in detail the total system costs of material, installation, and maintenance, and not just the initial product purchase price. For example, in many case, metal *plumbing* valves could be more than half the price of an injection molded plastic valve; however, when you consider joint integrity, the ease of joining, and lack of short- and long-term maintenance, the total cost savings of plastic valves are tough to beat. In chemically aggressive industrial applications, there are added cost savings using plastic valves because exotic alloy metal valves are normally much more expensive and, in many cases, require delivery times in weeks compared to the off-the-shelf availability of plastic valves.

Figure 5 Plastic flow meter.

Without the proper valve selections, no piping system will provide durable and cost-effective performance. Next time you witness metal valves in a plastic piping system, make it a point to educate the project's engineer, installer, and/or end-user to the many features and benefits of plastic valves.

Reprinted with permission of the IAPD; issue october/november 2009 – **the IAPD magazine**

CONCLUSION

Are plastics the panacea for all applications? No! But when conditions of service fit the scope of plastic materials, it would be wasteful and a disservice to the consumer and the planet not to consider the many pros of plastics, especially plastic piping systems.

Already, plastic piping is one of the preferred piping materials in such applications as: municipal water and sewer lines, drain-waste-vent lines, swimming pools, chemical waste drainage, aquariums, water and theme parks, rain harvesting, sprinkler systems, radon venting, central vacuum systems, double containment systems, residential fire sprinkler systems, trenchless underground piping, intake and outflow lines in power and desalination plants, ultra-high purity water systems, landfill gases and seepage, fish farms and hatcheries, mining fluids and wastes, surface finishing, natural gas distribution, residential hot-and-cold water distribution, and water/waste treatment.

Why the preference to plastics? It's simple; plastics are durable, easy and safe to install, environmentally sound, and cost-effective. These are the same reasons why any durable product should be selected as the material of choice.

I firmly believe that by the market place continuing to embrace the many features and benefits plastics have to offer, our planet and the future generations to follow will experience an environment that is cleaner, less wasteful, healthier, and more sustainable.

I hope you learn something useful from the articles and that you will be in agreement with me and the older character who, in the movie *The Graduate*, offered advice to young Dustin Hoffman as he was trying to get direction in his life. The wise man's advice was, "I just want to say one word...*plastics!*"

INDEX

www.ingramcontent.com/pod-product-compliance
Lightning Source LLC
Chambersburg PA
CBHW060327220326
41598CB00023B/2634